D1488551

SCHRÖDINGER
Centenary celebration of a polymath

SCHRÖDINGER

Centenary celebration of a polymath

Edited by

C.W. KILMISTER

Emeritus Professor of Mathematics, University of London

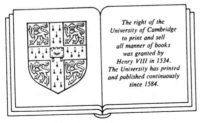

The right of the
University of Cambridge
to print and sell
all manner of books
was granted by
Henry VIII in 1534.
The University has printed
and published continuously
since 1584.

Cambridge University Press

Cambridge

New York New Rochelle

Melbourne Sydney

Published by the Press Syndicate of the University of Cambridge
The Pitt Building, Trumpington Street, Cambridge CB2 1RP
32 East 57th Street, New York, NY 10022, USA
10 Stamford Road, Oakleigh, Melbourne 3166, Australia

First published 1987
First paperback edition 1989

Printed in Great Britain at the University Press, Cambridge

British Library cataloguing in publication data
Schrödinger: centenary celebration of a
polymath
1. Schrödinger, Erwin
I. Kilmister, C.W.
509'.2'4   Q143.S3/

Library of Congress cataloguing in publication data
Schrödinger, centenary celebration of a polymath.
Papers presented at a conference, March 31–Apr. 3,
1987, at Imperial College, London.
1. Schrödinger, Erwin, 1887–1961 – Anniversaries, etc.
– Congresses. 2. Physics – History – Congresses.
3. Chemistry, Physical and theoretical – History –
Congresses. I. Kilmister, C. W. (Clive William)
QC16.S265S36   1987   530'.092'4   86-28338

ISBN 0 521 34017 9 hard covers
ISBN 0 521 37929 6 paperback

MC

Contents

Contributors

J. S. Bell
CERN-TH, 1211 Geneva 23, Switzerland

A. D. Buckingham
University Chemical Laboratory, Lensfield Road, Cambridge, UK

Jon Dorling
Goimburgwal 10, Gebouw 13, 1012 GA Amsterdam, The Netherlands

Dieter Flamm
Institut für Theoretische Physik, Technische Universität Wien, Austria

Kenichi Fukui
Kyoto Institute of Technology, Matsugasaki Hashigami-cho, Sakyo-ku, Kyoto 606, Japan, and Institute for Fundamental Chemistry, 15 Morimoto-cho, Shimogamo, Sakyo-ku, Kyoto 606, Japan

S. W. Hawking
Department of Applied Mathematics and Theoretical Physics, Silver Street, Cambridge CB3 9EW, UK

O. Hittmair
Institut für Theoretische Physik, Technische Universität Wien, Austria

Martin Karplus
Department of Chemistry, Harvard University, Cambridge, MA 02138, USA

T. W. B. Kibble
Blackett Laboratory, Imperial College, Prince Consort Road, London SW7 2BZ, UK

C. W. Kilmister
Department of History and Philosophy of Science, King's College London, Strand, London WC2R 2LS, UK

J. T. Lewis
School of Theoretical Physics, Dublin Institute for Advanced Studies, Dublin 4, Ireland

James McConnell
School of Theoretical Physics, Dublin Institute for Advanced Studies, Dublin 4, Ireland

Sir William McCrea

University of Sussex, Brighton, UK

Linus Pauling

Linus Pauling Institute of Science and Medicine, 440 Page Mill Road, Palo Alto, CA 94306, USA

M. F. Perutz

Laboratory of Molecular Biology, MRC Centre, Hills Road, Cambridge, UK

A. Salam

International Centre for Theoretical Physics, Miramere, POB 586, 34100 Trieste, Italy

M. J. Seaton

Department of Physics and Astronomy, University College London, Gower Street, London WC1E 6BT, UK

W. E. Thirring

Institut für Theoretische Physik, Technische Universität Wien, Austria

Chen Ning Yang

State University of New York, Stony Brook, NY 11794, USA

Preface

A conference was held from March 31 to April 3, 1987, at Imperial College London to celebrate the centenary of the birth of Erwin Schrödinger. It was supported by the Austrian Government, by the Dublin Institute for Advanced Studies and by many generous donations. A score of invited lecturers in the many scientific fields that Schrödinger had made his own were invited to survey his contributions and to discuss how it had influenced their own work. This volume contains most of these invited contributions.

1

Introduction

C.W. KILMISTER

University of London

This book describes the many aspects of the life and scientific work of Erwin Schrödinger, who, perhaps more than anyone else, serves to represent the whole of modern physics. The contributions to the book are from many hands, and so it seemed useful to me to precede them with a short note giving an uncontroversial account of his life which will serve as a framework into which they can be fitted. Such a collection as this still leaves out many aspects of Schrödinger's thought, for it considers his philosophy only in passing, his poetry not at all and ignores his wide and deep interest in sculpture and painting and in the classics.

Schrödinger was born in Vienna on August 12, 1887, and entered the University there to read physics in 1906. He worked there as an assistant from 1910 till his war service, and again after the war. Some short term appointments at Jena, Stuttgart and Breslau led up to his appointment to the chair of theoretical physics in Zürich in 1921. He was already treating a wide range of topics, but concentrating on atomic theory, for the old quantum theory had now entered on its heroic phase of final collapse. His six papers founding wave mechanics came at the end of his Zürich years, and in 1927 he went to the chair in Berlin, to remain there till the advent of Hitler in 1933. He chose to leave, though his own position could have been secure, and he spent a short and rather unhappy time in Oxford. But 1933 was also the year in which he shared the Nobel prize with Dirac, and he was offered a chair in Graz in 1936. Believing that there was no real danger of an *Anschluss* he accepted, only to have to leave hastily in 1938 for Rome. In that year he had some preliminary discussions with Eamon de Valera, and in 1940 the Eire Government established the Dublin Institute for Advanced Studies with Schrödinger as Director of the School of Theoretical Physics. Fifteen fruitful years followed, and his last five years were spent in his native land, where he died at Alpbach (Tyrol) on January 4, 1961.

Such are the salient facts, but it is not so easy to describe the history of Schrödinger's ideas. As in his general intellectual outlook, so in science in particular, his interests were wide, and yet in many fields his contributions were deep. His earlier pre-war papers exhibited interest in electricity and magnetism in solids, in statistical mechanics and Brownian motion and in acoustics. After the war he was at once in the thick of general relativity, but statistical physics and dielectrics still interested him and the former was a continuing interest. By 1920 colour vision had become important to him, and in the following year he became involved deeply in the old quantum theory. Of course this laid the basis in his mind for his discovery in 1926 of his preferred form of the new quantum theory, wave mechanics. This form was originally congenial to him as a means of getting rid of the objectionable quantum jumps. That it did not was a disappointment; he jokingly said he would never had invented it if he had known. As well as an acute aesthetic objection to the jumps, he retained a continuing disagreement with his close friend Max Born about the interpretation of the new formalism and indeed with the Copenhagen view in general. None the less wave mechanics continued as almost his sole occupation till 1934 when non-linear electrodynamics began to occupy his attention giving rise to a good deal of work on non-linear optics when he reached Dublin. The years after 1935 showed a second flowering of his genius as great as that of 1926 but this time over a wider field. He began to concern himself with Eddington's later theories, and, although he was of the opinion that Eddington's *Fundamental Theory* was mistaken in detail, this concern led him into cosmology, into the need for a proper rationalisation of the ground between relativity theory and quantum mechanics and into the algebraic structures which Eddington used. And all this, in turn, directed him towards the unified theories based on an affine connection, which occupied much of his Dublin time. It was not so much that Eddington had earlier formulated the first of these but rather that Schrödinger came to interpret his own theory, as it developed, by means of a philosophy near to Eddington's neo-Kantian views.

But Dublin did much more for Schrödinger, for occasional series of public lectures were amongst the responsibilities of the Senior Professor, and, although these were in a tongue that was not sung at his cradle (to use Weyl's phrase), he was fluent in it, for his maternal grandmother was English. Four of these series are available in book form, and the most notable of these is also the most technical, *What is Life?* (1944). The sales of this have probably reached 100 000 by now, and the interest in it when first published was intense. There were many highly favourable reviews. The scientific opposition was mostly confined to Schrödinger's suggestion that living things were characterised by their feeding on 'negative entropy' to stave off their final thermodynamic equilibrium. (Later on Schrödinger added a note explaining that he did not mean to ignore the energy value of food.) D. H. Wilkinson considered this a 'shocking expression', and J. A. V. Butler called it 'picturesque if somewhat inadequate'. On the other hand, Popper saw it as clear enough to attack the characterisation on the grounds that the same was true of steam engines, and the anonymous reviewer in the *Times Literary Supplement* found it a 'fascinating suggestion'. The theological criticism was more heavyweight, some of it on the same point but mostly concerned at the 'anti-christian philosophical speculations' in the epilogue which Schrödinger had added to the delivered lectures, indicating 'that the author is as poor a philosoher as he is a good scientist'. Professor O'Rahilly had demonstrated, said the *Catholic Herald*, 'that the work of the Senior Professor of the Dublin Institute for Advanced Studies culminates in a completely inacceptable doctrine of pantheism'. In fact, Schrödinger's speculation about personal immortality was an analogy drawn from the continuity between the same person at different times, arguing 'In no case is there a loss of personal existence to deplore. Nor will there ever be.' This quietistic, vaguely Eastern position makes a good end to this account of an extraordinary man whose work is still so much alive.

In writing this short account I have found Scott's bibliography (Scott 1967) invaluable.

References

Eddington, Sir A. S. (1946) *Fundamental Theory*. Cambridge University Press
Scott, W. T. (1967) *Erwin Schrödinger: an introduction to his writings*. University of Massachusetts Press
Schrödinger, E. (1944) *What is Life? The Physical Aspect of the Living Cell*. Cambridge University Press

2

Boltzmann's influence on Schrödinger

DIETER FLAMM
Technische Universität Wien

2.1. Introduction

Having the privilege of writing this paper as a grandson of Boltzmann I apologize for not being a historian of science. A manifestation of Boltzmann's influence on Schrödinger is Schrödinger's enthusiastic quotation of Boltzmann's line of thought: 'His line of thought may be called my first love in science. No other has ever thus enraptured me or will ever do so again.' (Schrödinger, 1929; reprinted in Schrödinger, 1957, p. XII.) Even though Schrödinger had no personal contact with Boltzmann, his scientific education at the University of Vienna was in the tradition of Boltzmann. His thesis advisor and director of the second Institute for Experimental Physics at the University of Vienna, Franz Serafin Exner, was an ardent admirer of Boltzmann and so was Boltzmann's successor on the chair for theoretical physics, Friedrich Hasenöhrl. Schrödinger paid the most impressive tribute to his teacher Hasenöhrl when he received the Nobel Prize in 1933. He avowed that Hasenöhrl might have stood in his place had he not been killed in the first world war (Schrödinger, 1935, p. 87).

Later one of Schrödinger's main fields of interest was the application of Boltzmann's statistical methods which he called 'natural statistics' to various problems. And it was not an accident that a paper in this field 'On Einstein's Gas Theory' triggered the idea of wave mechanics (Schrödinger, 1926, 1984). As he said, he formed a picture of the gas which corresponds to the picture of the black-body radiation applying literally the wave theory of moving particles by de Broglie (1924, 1925) and Einstein (1923, 1924). I shall later come back to the genesis of wave mechanics. For the moment let me give an outlay how I plan to display the influence of Boltzmann on Schrödinger. First I shall consider in more detail Schrödinger's academic training and stress the importance of Hasenöhrl's lectures and seminars. I shall also briefly speak about the years Schrödinger spent at Exner's institute. Then I shall discuss some quotations from Schrödinger's papers. It will also be rewarding to compare Boltzmann's and Schrödinger's views about life and its evolution. Finally I shall briefly compare their philosophical writings and mention Sir Karl Popper's report about his discussion with Schrödinger in his book *Unended Quest. An Intellectual Autobiography* (Popper, 1974).

2.2 At the University of Vienna

In autumn 1906, shortly after Boltzmann's tragic death, Erwin Schrödinger enrolled as a student at the University of Vienna. In his first year at the University Schrödinger presumably did not attend a course in theoretical physics. In autumn 1907, however, Boltzmann's student and successor Fritz Hasenöhrl gave his inaugural lecture. Schrödinger writes in his autobiography that Hasenöhrl explained in clear and enthusiastic words the basic ideas of Boltzmann's life work. This description made a deep intellectual impression on him, from which his thinking never departed from there on. Nothing seemed more important to him than Boltzmann's intellectual discovery (*Erkenntnis*) in spite of Planck and Einstein (Schrödinger, 1985, p. 15).

In an autobiographical sketch provided to the Nobel Committee in 1933 he wrote about the course in theoretical physics: 'During my four years at the University in Vienna (1906–10) the young Fritz Hasenöhrl, who had just stepped into the chair of the unfortunate Boltzmann, had the greatest influence. In a cycle of eight semesters for five hours per week were presented in detail the advanced theories of mechanics as well as the eigenvalue problems of physics of continua – in the detail in which later they would become necessary for me' (Schrödinger, 1935, p. 86).

Schrödinger mentioned in particular those topics in theoretical physics which he applied later in his wave mechanics. Of course, Hamilton's principle of least action, which was a favourite topic of Boltzmann, was among them. But Hasenöhrl taught Schrödinger other topics as well, notably the theories of Boltzmann. From Schrödinger's last year of studies there are six notebooks written in Gabelsberger shorthand with headlines and names in script. Five of them display Hasenöhrl's lectures in the winter term 1909/10 on heat theory treating kinetic theory, statistical mechanics and thermodynamics as well as problem sessions (Proseminar as they are still called today). In particular the fifth notebook gives the application of the statistical methods on the theory of black-body radiation and the sixth notebook is on Hasenöhrl's lectures on optics in the summer term of 1910.

There are also two smaller notebooks on a Hasenöhrl seminar in winter 1910/11 which was on gas theory.

Schrödinger thesis advisor was Franz Serafin Exner. He was five years younger than Boltzmann and had enrolled at the University of Vienna just when Boltzmann became recognized as academic lecturer. With Josef Stefan, Josef Loschmid and Victor Lang among his professors, Exner was bound to become a supporter of atomism. Since 1881 he was professor at the Vienna University, and in 1891 he became full professor and successor of Loschmid. When Schrödinger made his thesis 'On the electric conductivity on the surface of insulators in humid air', Exner was the most prominent professor of experimental physics in Vienna, and a whole circle of collaborators and former students had formed around him. This circle held regular meetings to discuss topics ranging from science to arts (Karlik and Schmid, 1982). Exner also promoted the heritage of Boltzmann, the central proposition of which is that the universality of the second law of thermodynamics is a consequence for macroscopic physics of the very existence of atoms (Scott, 1967). Exner carried Boltzmann's statistical reasoning even further by questioning the concept of causality on the microscopic level and proposing that all laws of physics are of statistical origin. He published these ideas in a textbook (Exner, 1919) which Schrödinger quoted in his inaugural lecture in Zürich in 1922 (Schrödinger, 1962).

Schrödinger (1887–1961) heard about Boltzmann not only from his professors, but also from his senior colleagues who had heard Boltzmann lecture and had passed examinations given by him, as for instance my father Ludwig Flamm (1885–1964) and K. W. F. Kohlrausch (1884–1953). By chance, as Schrödinger wrote (Schrödinger, 1985, p. 19), in 1911 he got a position at Exner's institute and not with Hasenöhrl. In these years before the first world war he also witnessed the fight between the epistemological schools of Boltzmann's and Mach's followers. Mach was a positivist. According to him the aim of science was the most economic book-keeping of experimental facts. He advocated purely phenomenological theories and rejected atomism and Boltzmann's statistical interpretation of the second law of thermodynamics. Boltzmann on the other hand was a realist and a passionate proponent of atomism and of the unification of mechanics and thermodynamics by statistical mechanics.

Altogether, Vienna, in 1911 when Schrödinger started his career as a scientist, was a stronghold of the followers of Boltzmann. At the Vienna Institute for Radium Research experimental evidence for the atomic structure of matter had been gathered from the study of radioactivity and of fluctuations. But Mach's reputation was high as is witnessed by the fact that he had been proposed for the Nobel Prize in physics in 1910, 1911, 1912 and 1914 by E. Suess, F. Braun, H. A. Lorentz and W. Ostwald (Blackmore and Hentschel, 1985, pp. 73, 88, 95, 129), respectively. Among Mach's followers were F. Adler (1879–1960), F. Ehrenhaft (1879–1952) in Vienna, P. Franck (1884–1966), who became Einstein's successor in Prague, A. Lampa (1868–1938) in Prague and J. Petzoldt (1862–1929) in Berlin.

The continuation of the struggle between Boltzmann and Mach started with a paper by Planck (1909) entitled 'The Unity of the Physical World Picture'. Planck criticized Mach's assertion that the primary aim of science was economy of thought as well as Mach's rejection of atomism and of the statistical interpretation of the second law of thermodynamics. In Planck's opinion Mach's philosophy was an obstacle for the unity of physics and in particular for the unity between mechanics and thermodynamics.

It is remarkable that between 1885 and 1889 Planck himself was a 'Machist' and Mach in 1888 even mentioned Planck together with G. Helm, L. Lange and J. Popper-Lynkens in the preface to the second edition of his book on mechanics. Only in 1900 had Planck become a follower of Boltzmann when he had to use Boltzmann's statistical methods for the theoretical derivation of his law of radiation (Planck, 1949). Mach's first reply to Planck was published in 1909 in the preface to the second edition of his book on the conservation of work (Mach, 1909). In reply Planck wrote a second paper called 'The Guidelines of my Scientific Knowledge: a Reply' (Planck, 1910b, c). In his dispute with Planck, Mach found support in two letters by Einstein (Blackmore and Hentschel, 1985, pp. 109, 121). A letter by Lampa (Blackmore and Hentschel, 1985, p. 80) shows that Mach and his followers speculated that the theory of relativity constituted the prelude of an epoch of phenomenology in physics.

In 1912 Mach's followers founded an association and a journal for positivistic philosophy. In later years Schrödinger himself defended Boltzmann's and Planck's point of view against the positivists. I should also mention Marian von Smoluchowski, who was educated at the University of Vienna from 1890 to 1894 and considered himself a student of Boltzmann. From 1906 to 1917 he worked out theories of Brownian motion and fluctuations which are macroscopic consequences of the very existence of atoms which could not be derived from phenomenological theories, but followed from Boltzmann's statistical theory. Smoluchowski's work was of great influence on Schrödinger. In fact, Schrödinger's own work on Brownian motion and fluctuations is closely related to Smoluchowski's papers. In particular in two handwritten notebooks

Schrödinger gives a detailed discussion of Smoluchowski's last papers. The transcription which amounts to 27 printed pages, is reproduced in an appendix of Hanle's thesis (Hanle, 1979, pp. 265–91).

2.3. Some hints in Schrödinger's papers

Boltzmann's work was the point of departure in many of Schrödinger's papers. I only have time to quote very few of them. For instance, Schrödinger's second paper 'On the Kinetic Theory of Magnetism (Influence of the Conduction Electrons)' (Schrödinger, 1984, p. 3), starts from Langevin's kinetic theory of magnetism and from the Lorentz electron gas in metals. For the proof that a magnetostatic field does not alter the velocity distribution of the free electrons in the metal he used Boltzmann's treatment of the velocity distribution for a mixture of two gases in the presence of arbitrary external forces.

Schrödinger's eighth paper 'On the Dynamics of Elastically Coupled Point-Systems' (Schrödinger, 1984, p. 124) is even more closely related to papers of Boltzmann in 1897 and to Boltzmann's article 'On the Indispensability of Atomism in Natural Science' (Boltzmann, 1905, p. 141; McGuinness, 1974). It treated the relation between the atomistic structure of matter and the phenomenological continuum theories of matter which are given by differential equations. Mach and the energeticists took the attitude that the phenomenological theories makes atomism obsolete. They called atomism a metaphysical concept which should be excluded from science. Boltzmann vigorously opposed this attitude. But to prove the indispensability of atomism he had not only to show that the phenomenological theories follow from atomism but moreover that atomism gives results of its own which cannot be explained from the phenomenological theories. Following Boltzmann's line of thought, Schrödinger wrote 'It is necessary to look for and to predict conditions under which the concept of a continuum, which rests on differential equations because of the truly atomistic structure of matters leads to observably incorrect results' (Schrödinger, 1984, p. 125).

The two quantities he considered are the calculation of the specific heat of solids and the temperature effect in the interference of X-rays. He studied a one-dimensional system and considered the relation between a vibrating chain, given by difference equations, and a vibrating string described by the one-dimensional wave equation. He brought the difference equations into the form

$$\frac{dx_n}{dt} = -\frac{v}{2}(x_{n+1} - x_{n-1}), \qquad (n = -\infty \cdots +\infty)$$

which is the well-known recursion relation for cylindrical functions x_n of argument vt. This enabled him to express the general solution of these equations as an infinite sum of Bessel functions of the first kind

$$x_n = \sum_{k=-\infty}^{\infty} x_k^0 J_{n-k}(vt).$$

In the following discussion he showed how the existence of atomic structure accounted for propagation of energy through the vibrating medium which is slow compared to the velocity of sound. In a second paper 'Dynamics of a Line of Elastically Coupled Points' he gave further applications of his method of solution (Schrödinger, 1984, p. 143).

Between 1914 and 1926 Schrödinger wrote eight papers related to the specific heats of solids and gases. Two of them are review articles. A thorough discussion of Schrödinger's papers in statistical mechanics between 1912 and 1925 is given by Hanle (1979) in his PhD thesis. Let me just mention Schrödinger's detailed *treatment of the deviation* of the mean energy for each degree of freedom from Boltzmann's equipartition theorem in solids in his review articles and his paper 'Remarks on the Statistical Definition of Entropy for an Ideal Gas' (Schrödinger, 1925, 1984, p. 341). In the latter paper Schrödinger compared various definitions of the entropy and discussed the division of the classical result for the number of microstates by the number of permutations of all molecules as proposed by Planck. Schrödinger corresponded with Einstein on the question of whether one should quantize the energy of a gas as a whole or consider the kinetic energy of every single molecule. Their correspondence led to the paper 'The Energy Levels of the Ideal One-Atomic Gas-Model' (Schrödinger, 1984, p. 143).

Now let us come to the roots of wave mechanics. In April 1926 Schrödinger wrote to Einstein (Przibram, 1963) that he might not have found wave mechanics if Einstein in his second paper on gas degeneracy (Einstein, 1923) had not drawn his attention to the importance of de Broglie's ideas (de Broglie, 1924, 1925). Schrödinger's motivation to work on Einstein's gas theory was to explain the new statistics which Bose had proposed for the quanta of radiation and which Einstein had applied to the atoms of an ideal gas in terms of Boltzmann's 'natural' statistical methods. In his paper on Einstein's gas theory (Schrödinger, 1926) he wrote

'The transition from natural to Bose statistics can always be replaced by an exchange of the parts, the manifold of energy states and the manifold of the carriers of these states play ... This implies to take serious de Broglie's and Einstein's undulatory theory of moving particles.'

Thus the heritage of Boltzmann gave the impetus for Schrödinger's work on wave mechanics. This is remarkable because Planck had also been motivated by Boltzmann to apply the methods of statistical mechanics to the theory of black-body radiation, which then led to Planck's quantization hypotheses. In 1897 Boltzmann had written

'It is certainly possible and would be rewarding also to deduce the radiation processes from a general principle which would be equivalent to the entropy principle for the theory of gases. I should be pleased if Planck's considerations on the scattering of electric waves by small resonators could be rendered useful in this direction' (Boltzmann, 1897).

Furthermore Schrödinger's book, *Statistical Thermodynamics* (Schrödinger, 1946), where he presented in expanded version many discussions of his paper 'On the Einstein Gas Theory' may be called a modern extension of the methods of Boltzmann.

Together with his friend K. W. F. Kohlrausch, Schrödinger published a paper called 'The Ehrnefest Model of the *H*-Curve' (Kohlrausch and Schrödinger, 1984). For this paper the authors actually performed Ehrenfest's urn experiment to study in detail the approach to equilibrium of Boltzmann's *H*-function. In his paper 'The Statistical Law in Nature' (Schrödinger, 1944, 1984, p. 451) Schrödinger paid tribute to Boltzmann on the 100th anniversary of his birth. A key problem of Boltzmann's *H*-theorem is whether one can arrive at irreversible behaviour starting from a reversible model. Schrödinger discussed it in his paper 'Irreversibility' (Schrödinger, 1950, 1984, p. 485). He stated his intention in the following form: 'I wish to reformulate the laws of phenomenological irreversibility ... in such a way that the logical contradiction which any derivation of these laws from reversible models seems to involve is removed once and for ever.'

2.4. Books and philosophical essays

Schrödinger and Boltzmann were both admirers of Darwin. In 1886 Boltzmann wrote with respect to the 19th century:

'If you ask me for my innermost conviction whether it will one day be called the century of iron, or steam, or electricity, I answer without qualms that it will be named the century of the mechanical view of nature, the century of Darwin.' (Boltzmann, 1905, p. 28; McGuinness, 1974, p. 15.)

In his essay 'The Spirit of Science' Schrödinger (1956) ranked Darwinism and Boltzmann's development of the statistical-mechanical theory of heat as the two leading ideas of 19th century natural science. After Boltzmann had calculated the temperature dependence of the entropy of radiation in 1884 he wrote

'The intermediate forms assumed by solar energy, until it falls to terrestrial temperatures, can be fairly improbable, so that we can easily use the transition of heat from sun to earth for the performance of work ... The general struggle for existence of animate beings is therefore not a struggle for raw materials ... nor for energy, which exists in plenty in any body in the form of heat, but a struggle for entropy.' (Boltzmann, 1905, p. 40; McGuinness, 1974, p. 24.)

The corresponding phrase in Schrödinger's book *What is Life?* (Schrödinger, 1948, p. 72) sounds very similar:

'What an organism feeds upon is negative entropy. Or, to put it less paradoxically, the essential thing in metabolism is that the organism succeeds in freeing itself from all the entropy it cannot help producing while alive.'

In a footnote to chapter VI of *What is Life?* he later explained why he did not use the concept of free energy:

'whose relation to Boltzmann's order-disorder principle is less easy to trace than for entropy and "entropy taken with a negative sign", which by the way is not my invention. It happens to be precisely the thing on which Boltzmann's original argument turned.' (Schrödinger's 1948, p. 86.)

For his book *What is Life?* Schrödinger had of course much more scientific evidence at his disposal than Boltzmann did nearly 60 years earlier, in particular he had quantum mechanics. The motivation for both of them to write about life was the universal aspect of science, as Schrödinger said apologetically in the preface of his book.

In his copy of Boltzmann's *Populäre Schriften* (Boltzmann, 1905) Schrödinger marked four articles entitled 'On the Indispensability of Atomism', 'More on Atomism', 'On the Development of the Methods of Theoretical Physics in Recent Times', and 'On Statistical Mechanics'.

Furthermore there are underlinings and a question mark in the text of 'On the Question of the Objective Existence of Processes in Inanimate Nature'. Here Boltzmann discussed idealism, realism and the scientific world picture. In particular he suggested that we should think of a machine by which everything we are empirically given of the mental processes would be realized. The following sentence has been underlined by Schrödinger; it reads: 'The rest we arbitrarily add in thought or so it seems to me.' But it seems that Schrödinger hesitated to follow Boltzmann's realism all the way, because he put a question mark next to the following sentence: 'Processes in inanimate nature differ so little in quality from animate ones, that it is impossible to draw any boundary.' (Boltzmann, 1905, pp. 184,

185; McGuinness, 1974, p. 23.) Karl Popper reported in his book *Unended Quest* about his discussions with Schrödinger. He wrote that he was surprised to find that Schrödinger was an idealist in spite of his greatest admiration for Boltzmann, who had been very critical about idealism (Popper, 1974, 1984, p. 195).

Another sentence Schrödinger has underlined in *Populäre Schriften* reads: 'This is the law of causality, which we are thus free to denote either as the precondition of all experience or as itself an experience we have in conjuction with every other.' (Boltzmann, 1905, p. 163; McGuinness, 1974, p. 75.) The corresponding passage in Schrödinger's inaugural lecture in Zürich in 1922 reads: 'This postulate is sometimes called the principle of causality. Our belief in it has been steadily confirmed again and again by the progressive discovery of causes that specially condition each event.' (Schrödinger, 1957, 1962, p. 10.) But in 1930 after the statistical interpretation of quantum mechanics he said cautiously: 'Currently it looks as if even the more modest desire for a picture which is at least in principle deterministic goes too far.' (Schrödinger, 1962, p. 24.)

Schrödinger's article, 'The Pecularity of the world-picture of natural Science' in 1947 demonstrates his obligation to Boltzmann's line of thought as he attacks therein Mach's extreme point of view and posivitism. Also in the spirit of Boltzmann is the headline: 'The picture is not only an allowed auxiliary means [of description] but also a purpose [of science]' in the same article. Irreversibility is another issue where Schrödinger followed Boltzmann's reasoning.

In his second reply to Zermelo, Boltzmann made the speculative conjecture that the direction of time might be defined by the direction of increasing entropy. For our expanding universe this is actually the case because, as it expands, its entropy increases with time. For a steady-state universe in thermal equilibrium there would, however, be no preferred direction of time. Life could only exist in subsections of such a universe which corresponds to thermodynamic fluctuations away from equilibrium. Boltzmann called these subsections *Einzelwelten*, and since fluctuations have a branch of deviation from equilibrium to locally smaller or larger entropy and a branch of return to equilibrium, he speculated that there might be Einzelwelten in which time goes in one direction, and others where time goes in the other direction.

Schrödinger (1950) followed this conjecture of Boltzmann for the definition of the arrow of time in his paper 'Irreversibility' to avoid any logical contradiction between the time reversal invariant laws of elementary particle physics and the irreversibility of thermodynamics. This point of view has been sharply criticized by Popper. He rejected it as a subjectivistic interpretation of the arrow of time which reduces the Second Law of Thermodynamics to a tautology (Popper, 1974, 1984, pp. 197, 230–6). On the occasion of the 75th anniversary of Boltzmann's death, M. Curd discussed Popper's critique in detail (Curd, 1982). Because of lack of time I shall only make a few comments. Boltzmann's original derivation of the *H*-theorem starts from his kinetic equation for the distribution of gas molecules in phase space, which is not time reversal invariant. To obtain it one has to assume that the velocities of the gas molecules are uncorrelated (*Stosszahlansatz*) and an average has to be taken. The challenge for mathematical physics is to actually derive this equation from classical dynamics. Lanford (1975), using the so-called Grad limit, has been quite successful in this respect. The Grad limit implies that the number of molecules in a finite volume goes to infinity retaining a finite mean free path. Thus in this limit a preferred direction of time is obtained starting from a time reversal invariant system. One may object that all really existing systems contain a finite number of molecules and thus irreversibility would be just an illusion (see, for example, Flamm, 1983). I believe, however, that the conceptual difficulty with the Grad limit is similar to the continuum limit of taking infinitely small volume elements in spite of the well-known atomic structure. It just works in most cases of macroscopic physics.

Acknowledgements

The author would like to thank Mrs R. Braunitzer, the daughter of Erwin Schrödinger, for her kind and spontaneous help looking up Schrödinger's personal library in Alpbach (Tyrol). He also thanks Dr W. Kerber, the chief librarian of the central physics library in Vienna, for his assistance and Dr A. Dick for valuable remarks.

References

Blackmore, J. and Hentschel, K. (1985) *Ernst Mach als Aussenseiter*. Wien: Braünmuller
Boltzmann, L. (1897) *S. B. Preuss. Akad. Wiss. Berlin*, p. 1016
Boltzmann, L. (1905) *Populäre Schriften*. Leipzig: J. A. Barth
Curd, M. (1982) in *Ludwig Boltzmann – Ausgewählte Abhandlungen, Ludwig Boltzmann Gesamtausgabe*, vol. 8 (Sexl, R. and Blackmore, J., eds), pp. 263–303. Graz: Akadem. Druck- u. Verlagsanstalt
de Broglie, L. (1924) *Thèses*. Paris: Edit. Masson & Cie
de Broglie, L. (1925) *Ann. de Physique* (*10*) 3, 22
Einstein, A. (1923) *S. B. Preuss. Akad. Wiss. Berlin*, p. 3

Einstein, A. (1924) *S. B. Preuss. Akad. Wiss.* Berlin, p. 261

Exner, F. (1919) *Vorlesungen über die Physikalischen Grundlagen der Naturwissenschaften.* Wien: F. Deuticke

Flamm, D. (1983) in *Entropy in the School,* vol. 1 (Marx, G., ed.), p. 11. Budapest: Roland Eötvös Physical Society

Hanle, P,. A. (1979) *Erwin Schrödinger's Statistical Mechanics, 1912–1915* (Yale University PhD Thesis 1975). Ann Arbor, London: University Microfilms International

Karlik, B. and Schmid, E. (1982) *Franz Serafin Exner und sein Kreis.* Wien: Österr. Akademie der Wissenschaften

Kohlrausch, K. W. F. and Schrödinger, E. (1984) in *Gesammelte Abhandlungen,* vol. 1 (Schrödinger, E., eds.), p. 349. Wien: Österr. Akademie der Wissenschaften

Lanford, O. E. (1975) in *Dynamical Systems, Theory and Applications* (Moser, J., ed.). Berlin: Springer

Mach, E. (1909) *Die Geschichte und die Wurzel des Satzes von der Erhaltung der Arbeit,* 2nd ed., p. IV. J. A. Barth, Leipzig

McGuinness, B., ed. (1974) *Ludwig Boltzmann – Theoretical Physics and Philosophical Problems.* Dordrecht: D. Reidel (English translation of Schrödinger (1984), p. 41)

Planck, M. (1909) *Phys. Zeitschrift* 10, 62–75

Planck, M. (1910a) *Phys. Zeitschrift* 11, 599–606

Planck, M. (1910b) *Phys. Zeitschrift* 11, 1186–90

Planck, M. (1910c) *Philosophie und Soziologie* 34, 497–507

Planck, M. (1949) *Vorträge und Erinnerungen.* Stuttgart: S. Hirzel, p. 25

Popper, K. R. (1974) *Unended Quest. An Intellectual Autobiography.* London and Glasgow: Fontana/Collins

Popper, K. R. (1984) *Ausgangspunkte.* Hamburg: Hofman and Campe

Przibram, K. (1963) *Schrödinger-Planck-Einstein-Lorentz – Briefe zur Wellenmechanik,* p. 24. Wien: Springer

Compare also Klein, M. J. (1964) 'Einstein and the Wave Particle Duality', *The Natural Philosopher* 3, 3–49. In a letter of November 3, 1925, Schrödinger wrote to Einstein: 'The de Broglie interpretation of the quantum rules seems to me to be related in some way to my note in the *Zs. f. Phys.* 12, 13, 1922 ... Naturally de Broglie's consideration in the framework of his large theory is altogether of far greater value than my single statement, which I did not know what to make of at first.' See Hanle, P. A. (1977) *ISIS* 68, no. 244, 606.

Schrödinger, E. (1925) *S. B. Preuss, Akad. Wiss.* Berlin, p. 434

Schrödinger, E. (1926) *Phys. Zeitschrift* 27, 95

Schrödinger, E. (1929) *S. B. Preuss. Akad. Wiss.* Berlin, pp. C–C11

Schrödinger, E. (1935) Autobiographical sketch in *Les Prix Nobel en 1933.* Stockholm: Imprimerie Royale

Schrödinger, E. (1944) *Nature* 153, 704

Schrödinger, E. (1946) *Statistical Thermodynamics.* Cambridge University Press

Schrödinger, E. (1948) *What is Life?* Cambridge University Press

Schrödinger, E. (1950) *Proc. Royal Irish Academy* 53A, 189

Schrödinger, E. (1956) *What is Life? and Other Essays.* New York: Doubleday Anchor Books

Schrödinger, E. (1957) *Science, Theory and Man.* New York: Dover

Schrödinger, E. (1962) *Was ist ein Naturgesetz?*, p. 14. Munich: R. Oldenburg

Schrödinger, E., ed. (1984) *Gesammelte Abhandlungen*, vol. 1. Wien: Österr. Akademie der Wissenschaften

Schrödinger, E. (1985) *Mein Leben – meine Weltansicht.* Wien: Paul Zsolnay

Scott, W. T. (1967) *Erwin Schrödinger, An Introduction to his Writings*, p. 16. University of Massachusetts Press

3

Schrödinger's original interpretation of the Schrödinger equation: a rescue attempt

JON DORLING

University of Amsterdam

Schrödinger's original interpretation of the Schrödinger equation had many attractive features lost in later interpretations of the quantum theory. But that interpretation runs into a number of formidable well-known and not so well-known objections. I argue, following the methodological precepts of Paul Feyerabend, that we need not regard any of these objections as fatal, provided we are prepared to opt for a number of bold and rather radical mathematical and theoretical conjectures. These would amount jointly to the conjecture that a fully time-symmetric consistently classically interpreted non-second-quantized analogue of existing quantum field theory would (*pace* Jaynes, Tomonaga, Bell, and others) ultimately prove predictively equivalent to orthodox second-quantized theory.

Introduction

Schrödinger initially proposed his equation as a classical theory of matter waves directly analogous to Maxwell's theory of electromagnetic waves. $|\psi|^2$ represented a classical charge density functioning in the ordinary classical way as a source of electromagnetic fields, and acted on by these fields via the potential term in the matter–wave equation. This is a theory of coupled classical fields with no probabilities entering into its interpretation, and from a modern point of view it can be thought of in terms of the coupled Dirac and Maxwell fields, without second-quantization and interpreted in a purely classical manner. Such a theory in fact generates the usual atomic wave-functions* and explains the frequencies, relative intensities, and polarizations of spectral lines, and the

---

* See chapter 6 of Tomonaga (1966) for a consistently classical presentation of this theory and its applications to nearly all the standard problems of non-relativistic quantum mechanics.

associated selection rules, in terms of the radiation which would be classically emitted by the charge cloud oscillations which now calculably arise in superposition states*.

This original theory of Schrödinger's met standards of objectivity, realism, and physical picturability which have not been attained by any of the versions of the quantum theory which succeeded it. As in the case of any provisionally discarded theory which possessed attractive features lacked by its immediate successors, the historian and philosopher of physics must ask whether there is still a possibility that one day in the future such a theory may be revived and reinstated. For not all the original objections to a discarded theory generally stand the test of time, and such objections, together with any later ones, need to be periodically re-assessed in the light of subsequent progress in physical theory.

My interest is not so much in whether Schrödinger's original interpretation of his equation is still tenable, but rather in what it would be necessary to believe in order to consider it as still tenable. Given the empirical success of the orthodox second-quantized theory, one would have to believe something like that this second-quantized theory would ultimately turn out to be merely a conservative extension of classically interpreted first-quantized theory. This may seem a very remote possibility, but until we really understand what a purely first-quantized world would be like, and precisely how it would differ from our actual second-quantized world, it is a possibility which cannot be dismissed without careful investigation†. My own hunch here is that contemporary theorists are a little too generous in admitting alternative mathematically possible worlds different from our actual physical world, and that it could be that all alternatives would ultimately turn out either to be incoherent, and thus not to describe possible worlds at all, or turn out, if consistently carried through, to be merely redescriptions, in a different theoretical vocabulary, of our actual physical world. That is to say, maybe a clever enough theorist could in fact save all the phenomena from any one of a considerable diversity of apparently mutually incompatible theoretical starting-points. Schrödinger's original viewpoint would be merely one such permissible starting-point. What I want to suggest is that it was something of a

---

* See chapter 33 of d'Abro (1939, 1951) for a clear informal account of this aspect of Schrödinger's interpretation.
† E. T. Jaynes's (1973) strategy in his work on 'neo-classical radiation theory' was rather to seek here a rival to second-quantized theory and to look for crucial experiments. My strategy is rather to suppose that a fully coherently formulated first-quantized theory, if such is possible at all, ought, if we do everything correctly, to turn out predictively equivalent to second-quantized theory.

historical accident that Schrödinger's original programme for the interpretation of the quantum theory was abandoned so hastily, and that, had the tastes of theoreticians been different, it could, at a certain price, have been retained. I want to spell out what that price might be, just what expectations a theorist would have to have were he to suppose that Schrödinger's original interpretation of the quantum theory will one day be reinstated.

Specifically, if only as a constructive heuristic exercise in envisaging possible escape routes from current physical orthodoxy, I would like to explore the following suggestions for the would-be reviver of Schrödinger's viewpoint.

(1) The problem of Schrödinger's matter waves being defined for an n-particle system on a $3n$-dimensional space, rather than on ordinary physical 3-dimensional space. A possible way out here would be to attempt to generalize Feynman's device of treating certain many-particle problems by means of a non-second-quantized one-particle theory. This device, which interprets pair-creation and annihilation as involving a single particle scattered backwards and forwards in time, would need extending to many-particle problems in general. To cope with particles of different sorts we would need to generalize this one-field approach further to the context of non-second-quantized grand unified theory.

(2) The problem (especially emphasized by Tomonaga) that Schrödinger's non-probabilistic matter–wave theory really only explains the wave properties of matter, but leaves most of the particle properties unexplained, being ultimately no more able to explain mass and charge quantization, or the $E = h\nu$ relation between energies and frequencies of matter waves, than Maxwell's theory is able to explain the latter relation for electromagnetic waves. Here it seems that a possible way out is to argue that the non-linearity introduced into the classically interpreted Schrödinger or Dirac equation, by the self-coupling of the ψ-field via the classical electromagnetic field, actually makes it mathematically plausible that the missing quantization conditions may in principle be derivable within this theory.

(3) The problem of 'locality' for any interpretation like Schrödinger's which construes the quantum theory as ultimately just a special kind of classical relativistic field theory. Here I offer reasons for supposing that *any* time-symmetric classical relativistic field theory, provided we do not effectively cancel out all advanced causal effects by imposing arbitrary and empirically implausible cosmological boundary conditions, must be

mathematically expected to be non-local in the same sort of way that the quantum theory is non-local. So I claim to turn this objection on its head. (4) The problem of wave-packet collapse for Schrödinger's interpretation. Here again I suggest that the answer may merely lie in keeping our essentially classical theory fully time-symmetric. For in a causally time-symmetric classical relativistic theory an emitter does not transfer energy and momentum to the radiation field along an infinity of directions in space, but only along the null line(s) joining emitter and actual absorber(s). On such a view probabilities would enter a consistent Schrödinger interpretation of the quantum theory only via the experimeter's present ignorance of the relevant parameters associated with all the potential future absorbers.

My aim in the following will be the modest one of defending against obvious ridicule these and related suggestions for saving Schrödinger's original programme. For I wish to contend simply that the historian of physics cannot *yet* take it for granted, in the light of our *present* understanding of the foundations of the quantum theory, that Schrödinger's original viewpoint will not be reinstated at some time in the future. For we have not, I believe, yet found any entirely satisfactory interpretation of the quantum formalism, and it is therefore ultimately a matter of scientific taste which programmatic interpretation, together with its so far unresolved special difficulties and conjectural solutions to these, one opts for. Furthermore it does not seem to me likely that the final answer will prove to be one of the currently more fashionable interpretations, for were that to be the case, it is hard to see why the best theorists should, after 60 years of the quantum theory, still be so far from unanimity on the technical details of the correct interpretation of the theory. I am therefore inclined to think that the race to find a finally convincing interpretation of the quantum theory will eventually be won by a rank outsider. One possibility is of course that there is something fundamentally wrong with the conventional mathematical formalism of the quantum theory – perhaps we ought to be using only discrete mathematics, never introducing the continuum mathematics of classical field theory at all; but assuming that the conventional mathematical apparatus is indeed applicable, I think we need to understand a good deal more about what sort of physical world, if any, a classically interpreted non-second-quantized theory would describe, before we dismiss Schrödinger's original interpretation of the quantum theory as not merely an outsider but a non-starter.

3.1. The original objections: wave-packet dispersion, and the problem of Schrödinger's matter waves in $3n$-dimensional space

I shall argue that the first problem is really just as much a problem for the modern Copenhagen interpretation, and that even without second-quantization we can in principle handle n-particle problems without introducing waves in $3n$-dimensional space, by adopting and extending an ingenious trick of Feynman's for reducing such problems to one-particle problems. So I shall be arguing that the use of such waves in $3n$-dimensional space is merely a convenient calculational device which is in principle theoretically eliminable and so cannot be counted as an objection to the Schrödinger view. It is true that Feynman's actual proposal could only do the trick for a many-particle system containing only particles all of the same sort and their anti-particles. But this restriction seems to disappear in the context of a grand unified theory, as does the difficulty that Feynman's original trick seemed to require as many positrons as electrons (± 1) in the physical universe. So the Schrödinger interpretation with all fields defined (theoretically and in principle) only on ordinary physical space seems to have become a technically viable option again.

Of the two objections which were principally responsible for the early abandonment of Schrödinger's original viewpoint, the first was the failure of Schrödinger's attempted representation of particles by non-dispersive wave-packets. But I deny that this is any more or less of a problem for the Schrödinger interpretation than it is for any modern Copenhagen interpretation. What happened historically was that the immediate successor to Schrödinger's interpretation, namely the naive Born statistical ignorance interpretation of $|\psi|^2$, immediately dissolved this problem. But the naive Born interpretation is now universally abandoned, and its successors are again faced with Schrödinger's original problem. For it has become abundantly clear from analyses of the two-slit experiment, from the theorems of Gleason and of Kochen and Specker, from Cohen's no joint-probabilities theorem, and from Bell's theorem, that no naive statistical ignorance interpretation of the quantum mechanical wave-function will work. A modern Copenhagen interpretation is, in this respect, much closer to Schrödinger's original interpretation than to Born's original interpretation, and the problem which Schrödinger was here grappling with unsuccessfully is still with us, partly as the problem of wave-packet collapse, which I shall return to in section (4), partly as the problem of how one should theoretically derive cloud chamber tracks. The most satisfactory theoretical derivation of cloud chamber tracks, one not relying on Heisenberg's artificial device of introducing successive wave-packet collapses, or on the construction of non-dispersive wave-packets, is that given in section 30 of Schiff (1955). The incident electron is merely represented as a plane wave, and the essentials of Schiff's technical derivation go through equally well on either a Copenhagen or a Schrödinger interpretation of the wave-function. If one is not satisfied with this or related derivations of particle-like trajectories, then the Schrödinger interpretation merely shares a problem here with all the more modern orthodox interpretations. Hence this historically influential objection should no longer be cited as a significant special objection to Schrödinger's viewpoint.

Much more difficult is the second original objection to Schrödinger, namely that in the case of a many-particle system the Schrödinger waves apparently cease to be interpretable as matter waves in ordinary physical space.

Now I think we should first ask ourselves, is this conventional non-relativistic many-particle formalism more than a calculational approximation? In particular is it really correct in the relativistic case? It seems to require that the total number of particles of all sorts be strictly conserved. But a Dirac or (Schrödinger–)Klein–Gordon particle evolves even according to its one-particle equation, given an appropriate potential, into negative energy or positive charge states, and we seem to have to interpret such evolutions in terms of pair-production and pair-annihilation processes, in violation of the conservation law apparently built into the many-particle formalism here under discussion. Now of course one could take the view that there is no consistent relativistic first-quantized theory, and that the only solution here is to go over to the second-quantized theory. This was I think at one time the orthodox opinion. But it is an odd opinion, for one would think that, given a mathematically internally consistent physical theory, it is only us who are to blame if we have not yet found a consistent physical interpretation for it. And the mathematical internal consistency of relativistic first-quantized theory is surely a precondition for getting a mathematically consistent theory by second-quantizing it. From these reasonable premises it would follow that it must have a consistent physical interpretation even if we have not yet found it. This situation was considerably clarified by Feynman's contributions to quantum field theory of the late 1940s and early 1950s.

I think one can extract from these contributions of Feynman, particularly from Feynman (1950) and its appendixes, the following view. The so-called 'single-particle' Dirac or (Schrödinger)–Klein–Gordon equation, without second-quantization, but with an arbitrary electromagnetic potential, includes not only ordinary scattering, but also scattering backwards and forwards in time, which we interpret physically as pair-annihilation and pair-creation processes. Hence, such an equation implicitly covers the case where many particles and their associated antiparticles are present simultaneously in one reference frame (a single electron being scattered backwards in time as a positron, then forwards again as an apparent second electron, then . . . and so on). Hence, in principle one should be able to treat the problem of, say, n electrons interacting with one another, by the one-electron formalism, by supposing that $n-1$ pair-creation processes occurred in the remote past and $n-1$ pair-annihilation processes in the remote future, and by introducing suitable hypothetical electromagnetic fields in the remote past and the remote future to ensure that the $n-1$ positron trajectories linking these creation and annihilation processes remain wholly separate from the n electron trajectories for the bulk of the intervening period. Of course such a treatment of an n-electron problem by the one-electron formalism would be highly artificial and contrived, but it seems that in principle it ought to be predictively equivalent to the usual first- or indeed second-quantized many-particle formalism for the region of physical interest. In particular the symmetrization requirements of the usual many-particle formalism, that is in the Dirac case Fermi–Dirac and in the Klein–Gordon case Bose–Einstein, must it seems already be implicit in the mathematics of the corresponding one-particle theories. The point here is not that one should try to use the one-particle formalism in this way to treat the many-particle case, but rather that any possible many-particle physical scenario (involving only particles of the one sort and their corresponding antiparticles) could always turn out to have been merely a part of such a one-particle scenario; hence, all the physical theory of such many-particle situations must already be implicitly contained in the complete physical theory of the case of one particle in an arbitrary electromagnetic potential. But that means that for the physics of any such many-particle system, which we most conveniently treat using the formalism of fields defined on an abstract higher-dimensional space rather than on ordinary physical space, there must exist (theoretically and in principle) an empirically equivalent description only employing (non-second-quantized) fields defined on ordinary 3-dimensional physical space. Hence, in the restricted case of like particles and their corresponding antiparticles, the many-particle formalism, with its requirement that our fields are defined on $3n$-dimensional rather than on ordinary 3-dimensional space, is really theoretically redundant. We use that formalism merely as a practical convenience, but, since it is not theoretically essential in principle, its use here cannot in any way count against Schrödinger's position in which all fields must be ultimately regarded as defined on ordinary 3-dimensional space.

It is true that Feynman's series of papers never said quite this absolutely explicitly, and I have had largely to reconstruct such a position from remarks which Feynman made as asides, as footnotes, and in appendices, or where he presented plainly programmatic guesses (which I have sometimes taken the liberty of generalizing). And of course he never mentioned the Schrödinger interpretation as such. However, the point is this. If we consider any relativistic analogue of the Schrödinger equation with an arbitrary electromagnetic potential included, we can no longer interpret such an equation as defining a conventional one-particle theory: positive charge or negative energy states inevitably arise as a result of the action of the potential, and one must choose between either denying that there is any consistent physical interpretation of such a theory, or consistently reinterpreting it physically as a one-field but many-particle theory, including pair-creation and annihilation processes, and in principle including situations where many particles and their associated antiparticles are simultaneously present in one reference frame. This seems to me inescapable. However, this approach, as I have sketched it, still only solves a restricted special case of the problem here confronting the Schrödinger interpretation, for how on earth are we to treat a problem involving several *unlike* particles, for example the proton and the electron forming a hydrogen atom, as a Feynman one-particle problem?

First I need to make a remark concerning the recent gauge theoretical revolution in particle physics. This is really an advance in classical field theory. It is true that the second-quantized versions of gauge theories with spontaneous symmetry-breaking prove to be renormalizable, and that this is a pleasant bonus from the point of view of calculations within the orthodox second-quantized framework, but in fact all the real conceptual work of gauge theories, including the spontaneous symmetry-breaking, appears to take place at a classical field-theoretical level, independently of second-quantization. Peter Higgs has repeatedly emphasized that his contribution was essentially a contribution to classical physics. In particular it needs to be clearly understood that the explanatory power of the grand unified gauge theories, the unification they offer, and the way out they offer from the centuries-old concept of arbitrary forces or potentials in nature, is quite independent of second-quantization, and hence provides no evidence against Schrödinger's original classical field theory inspired programme; I shall suggest that the reverse is rather the case.

My proposal is that when, within a fully unified theory of the different particle species, one identifies different fermions (perhaps including combinations of them) as different states of a single underlying field, the path becomes open for applying the Feynman trick to *all* many-particle problems. Consider, for example, the proton and the electron in a hydrogen atom. We could treat this as a one-particle problem in either of two ways. We could trace the proton trajectory back in time until it curves violently and becomes an antiproton which then emits an exotic boson and appears as the electron. Or we could trace the electron trajectory back in time until it curves violently and becomes a positron which then absorbs an exotic boson and (by the inverse of proton decay) becomes the proton. In the first case we must say that the full one-particle equation for the proton, including all boson fields to which it is coupled, but without second-quantization, and allowing for scattering backwards and forwards in time, must implicitly include the full theory of the hydrogen atom. And on the second analysis we must say the same of the full one-particle theory of the electron. Given a Feynman-style interpretation of the non-second-quantized version of this kind of theory, it no longer seems possible to say that there is any problem which actually requires the many-particle rather than the one-particle formalism. And also, to carry out this trick, we no longer need to appeal to a fiction of more-or-less exact matter–antimatter symmetry in the universe.

So I say that when considering the objection to Schrödinger that Schrödinger-type waves have to be defined on a non-physical higher-dimensional space, we should be looking not at what is practically convenient for certain calculations, but at what is theoretically possible in principle. And it seems to me that there are reasons for supposing that all the usual many-particle situations ought theoretically in principle to be already covered at non-second-quantized level, by the full, correctly interpreted, one-particle equations, and hence that (although we should surely never wish to use such a treatment in practice) a field or fields defined merely on ordinary physical space would suffice in principle in all cases. And this would then eliminate the most notorious objection to Schrödinger's programme.

I am not claiming that I expect in these brief informal remarks to have eliminated this objection to anyone's satisfaction. I am merely arguing that there are subtle and delicate theoretical issues involved which involve theoretical taste and theoretical judgement and that we cannot thus without more ado consider the existence of many-particle problems, and the approximate correctness in its predictions of their conventional non-relativistic treatment, as any longer a knock-down argument against Schrödinger. At any rate, *I* do not know enough to exclude Schrödinger's interpretation in this way. I am merely arguing for a certain open-mindedness, as a result of recent theoretical developments, on what has hitherto been treated as a closed issue. Do we really know? I don't think we do yet. And our guesses about future theoretical developments depend psychologically on what we would like to be the case. What Schrödinger would have liked to be the case is clear.

3.2. Is a first-quantized theory really quantized at all?

A much more radical objection to Schrödinger's programmatic theory was clearly identified and discussed only, to my knowledge, many decades later, in S. I. Tomonaga's textbook (English edition, Tomonaga (1966)). The objection is this: given that Maxwell's field theory of radiation cannot explain the quantization of the energy of electromagnetic waves, how can it be that Schrödinger's analogous theory of the matter field can explain the quantization of the energy of matter waves? Tomonaga argues that in fact it cannot, that, although this theory indeed yields discrete frequency spectra for the matter waves, it contains no mechanism for associating these with the corresponding discrete energy spectra: $E = h\nu$ is no more a theorem in Schrödinger's theory of matter waves than it is a theorem in Maxwell's theory of electromagnetic waves. 'First-quantization' is thus a misnomer; Schrödinger's theory of electron-waves, as a replacement for the particular theory of electrons, no more involves quantization than does the Young–Fresnel–Maxwell theory of light-waves as a replacement for the Newtonian theory of light-corpuscles.

How are we to escape from this objection? At first it seems that Tomonaga cannot be right, because Planck's constant seems to make a very definite appearance in the Schrödinger and Dirac equations for the matter waves, whereas it makes no appearance in Maxwell's equations for the electromagnetic waves. However, Tomonaga argues in considerable detail that, from the point of view of a consistent Schrödinger interpretation of the Schrödinger equation, the appearance of Planck's constant in that equation is a formal illusion. For from the point of view of a consistent wave-interpretation we must consider the potential function divided by Planck's constant as a refractive index function for the matter waves, and it is only this refractive index function which is directly empirically interpretable from the point of view of the Schrödinger interpretation. But this only leaves Planck's constant in the equation in the combination m/h, a constant which relativistically has the dimensions of a frequency, and which is not further analysable without begging the question of mass–energy quantization. Solution of the Schrödinger equation for a specified refractive index function yields stationary matter–wave solutions with calculable characteristic frequencies. But the energy associated with such a frequency remains arbitrary, and indeed, because of the linearity of the equation, indeterminate. For the absolute amplitude of the matter waves can, because of this linearity, take arbitrary values, and the total energy is only determined as an integral over these amplitudes over all space. The problem is that the usual normalization condition on the wave-function not merely has no justification from the point of view of a consistent wave-interpretation, but is not, if we are careful with units, even formulable without begging the question of mass, charge and energy quantization.

I think that Tomonaga's considerations here are basically correct. From the point of view of the coupled Maxwell and Dirac fields without second-quantization, we find the constants that we can assume specified are a constant of the form m/h giving the intrinsic rest-frequency of the electronic matter field, and a coupling constant e^2/h coming from a constant e in the source term for the electromagnetic field and a constant e/h in the Dirac equation. But only the product of the values of these latter two constants, i.e. the effective coupling constant, is significant. Thus the charge e, the mass m and Planck's constant h make no separate appearance in this system of equations. While the total charge and total mass–energy will vary proportionately to each other, neither is *prima facie* either calculable or a permissible primitive ingredient in the theory.

This seems indeed the deepest objection to Schrödinger's original interpretative programme. In particular these considerations present a formidable undiscussed difficulty for Jaynes's attempt to revive that programme in conjunction with his 'neo-classical radiation theory' (e.g. Jaynes (1973)). Jaynes's idea was that it was not in principle necessary to quantize the radiation field, for provided the interaction of atoms and the radiation field were such that radiation could only be emitted or absorbed in discrete amounts determined by the quantum levels of the atoms, independent quantization of the radiation field would be unnecessary, and we might as well continue to treat it using ordinary Maxwellian theory. In order to obtain consistently coupled matter and radiation fields, this requires, as Jaynes appreciated, that we must also reject second-quantization for the matter field and return to the original Schrödinger interpretation in that case. But the difficulty is that we now see, in the light of Tomonaga's considerations, that we then fail to get, among other things, the necessary energy quantization for the atomic states, without which it is simply no longer true that the radiation energy can only be emitted or absorbed in the correct discrete amounts.

We can also run this argument in reverse. If we stick to Schrödinger's first-quantized theory of the matter fields, we get inconsistencies if we try to couple this with a second-quantized theory of the radiation field. Hence we can only couple it with Maxwell's classical theory of the radiation field. But then how can Schrödinger's interpretation of the matter–wave field ever be extended to yield an explanation of the existence of photons?

We can see that this problem is intimately connected with the linearity of the Schrödinger or Dirac equation, in the following way. Consider the ground state of the hydrogen atom, with the proton represented to a good approximation by a Coulomb field, and with the electron wave-function representing, on the Schrödinger viewpoint, an extended charge cloud whose charge density and mass density are proportional to $|\psi|^2$. Since the Schrödinger equation is linear in ψ, if we increase the amplitude of ψ everywhere by the same proportional amount, then we still have a solution. But this will correspond to a greater total charge for the charge cloud and to a greater total mass and energy for the atom. Instead of one ground-state solution we get a continuum of solutions differing only in the normalization of ψ. For unless we have already opted for an interpretation of $|\psi|^2$ as determining not a physical charge and mass density but the probability of finding a particulate electron at the point in question, we can no more require that the integral of $|\psi|^2$ over all space be normalized to unity, than we can require of classical electromagnetic waves in a cavity that the total energy associated with any given frequency v be normalized to an integral multiple of hv.

I shall now defend a very radical solution to this problem. I deny that the Schrödinger and Dirac equations, correctly understood, are linear in ψ. First it is clear that from a consistent Schrödinger viewpoint one cannot simply ignore the additional contribution to the electromagnetic field acting at any point, which should be produced by the rest of the electron charge cloud. In fact one is theoretically required in the treatment of the hydrogen atom to add to the term in the wave equation representing the Coulomb field of the proton an additional term representing the effect of the rest of the electronic charge cloud on itself and which is evidently of the form

$$e^2 \iiint \frac{|\psi(\mathbf{r}')|^2}{|\mathbf{r} - \mathbf{r}'|}\, dv'.$$

But this of course turns the whole equation from a linear differential equation into a non-linear integro-differential equation in ψ.

This change is not only forced on us by the Schrödinger interpretation. Given any interpretation whatever of the first-quantized coupled equations for the matter and radiation fields, a charged matter field must act as a source term for the electromagnetic field, and the resultant self-coupling induced via the electromagnetic potential term in the wave equation for matter simply cannot be ignored on pain of contradiction. The conventional first-quantized treatment of the hydrogen atom with an equation linear in ψ must therefore be simply theoretically incorrect.

Tomonaga was aware of this problem, and his view was that such an extra non-linear contribution to the usual theory was indeed theoretically required, but that its addition would inevitably yield a new and empirically incorrect spectrum for the hydrogen atom, and that only second-quantization could resolve this impasse. But in fact Barut and Kraus (1983) have recently succeeded in showing that Tomonaga was wrong here, and that if one includes such a non-linear self-coupling term in the Dirac equation (including of course also the corresponding vector potential contributions) and then solves the resulting equation for the hydrogen atom by iteration, then (after carrying out the appropriate mass and charge renormalizations) two things happen: (i) the only changes yielded in the hydrogen spectrum correspond to the appearance of the Lamb shift and of the anomalous magnetic moment of the electron, (ii) the solutions other than the ground state solution are no longer strictly stationary but contain additional exponential decay terms corresponding to the probabilities of spontaneous emission. (This is all without ever introducing second-quantization.)

So it seems to me that we have today to accept the essential and radical non-linearity in ψ of the correctly applied Schrödinger and Dirac equations. But then this completely transforms the problem for Schrödinger's interpretation which we have been discussing in this section. For it is now no longer the case that given a solution to this equation (other than perhaps a free-particle solution) that the normalization of ψ (and with it the total charge, mass and energy associated with the charge cloud) can remain theoretically indeterminate. For now if we alter this normalization, we alter the 'potential' term in the wave equation, and our original solution is no longer an eigensolution, in contrast to the situation with the conventional, but theoretically incorrect, linear equation in ψ. Hence, total mass, charge and energy are no longer arbitrary, and this classically interpreted first-quantized theory must now yield not only a frequency spectrum, but also the correspondingly discrete energy spectrum. Indeed the value of Planck's constant, or equivalently that of the total electronic charge, must in principle be calculable within this theory.

What this, I think somewhat surprisingly means is that the value of the fine-structure constant ought to be calculable from the same equations if we give the matter equation the conventional particle interpretation. This is partly a matter of seeing that one can only calculate the values of dimensionless quantities. But it can be seen more directly as follows. There are two dimensionless quantities, the fine-structure constant and the normalization to *unity*. If one considers again the ground-state case, then taking the fine-structure constant as given, the normalization of ψ should be calculable because for the ground-state solution it is, because of the non-linearity, no longer arbitrary. But if by fiat we had fixed the normalization as normalization to unity, then the only free parameter left to vary so as to satisfy the new constraint arising from the non-linearity would be precisely the fine-structure constant. So on the Schrödinger interpretation we must assume the fine-structure constant as given and calculate the value of another dimensionless quantity as equal to unity. On the usual interpretation we would take the second constant as given and calculate another dimensionless constant equal to approximately 1/137. But both calculations are essentially the same, because the only place where changing the normalization makes a difference is where the normalization factor is multiplied by the fine-structure constant. (I ignore mere complications due to the possibility of non-integral charges.) This situation is at first confusing because we ordinarily consider the number one as unproblematic, but I think this is only when we consider quantization itself as unproblematic and not requiring explanation.

The suggestion that the value of Planck's constant, or equivalently that of the fine-structure constant, ought to be theoretically calculable is an old one. Three recent putative calculations within a framework related to that here under discussion are worth mentioning. Wyler (1969) 'derived' the fine-structure constant on the basis of abstract group-theoretical considerations concerning the Poincaré group and the Conformal group. Nobody has really understood these considerations, though these are of course the groups appropriate to the interacting matter field and radiation field in which we are here interested. Barut (1978) obtained values of Planck's constant (assuming merely the total electron charge as given) and of the fine-structure constant, from an analysis of the classical radiation which would be emitted by a Dirac electron, construed as a point charge undergoing a Schrödinger Zitterbewegung. His argumentation is not entirely rigorous and his results possess certain mysterious features. I am not clear whether the self-coupling of the charge with its own radiation field could be playing any implicit role in this calculation; its role is certainly not explicit in it. However, directly related to the present suggestion is the calculation by Samec (1980) of the value of the fine-structure constant. This is a direct attempt at a numerical computation of the charge-quantization constraint arising from the non-linear coupling between the Maxwell and Dirac fields which is here under discussion. Unfortunately, while Samec claims to have calculated a reasonable value of the fine-structure constant, he obtains a very mysterious spectrum of integral and third-integral charges which it is hard to take seriously physically. (Maybe Samec has simply made one or more technical mistakes, maybe it won't come out right unless you also put in all the non-linear self-couplings associated with the other gauge fields.)

What I would rather emphasize here is that a consistent Schrödinger viewpoint does seem to require that the value of Planck's constant, or equivalently that of the fine-structure constant, be calculable, and that it is by no means excluded that this may actually be possible within the theory generated by the non-second-quantized coupled Maxwell and Dirac fields. We do not need to be able to calculate such a value within orthodox second-quantized theory, since it is enough that it exists and has a definite value, since this is necessary for the formulation of that theory. However, if we stick to the first-quantized theory without smuggling in a particle interpretation, just as if we stick to Maxwell's theory of radiation, the very existence of Planck's constant becomes mysterious, and nothing short of its calculability (presumably on the basis of some overlooked non-linearity, for what else could render it calculable?) would seem to be capable of resolving the mystery. It is conceivable that some kind of cosmological theory of the value of the constant in question might also suffice. It has become fashionable recently to suppose that the value of the fine-structure constant is something that physicists never need to be able to calculate theoretically, for if it were not roughly what it is then the evolution of atoms beyond carbon would not have occurred and we would not be here. But such a view could never explain the *exact* value of the fine-structure constant, but at best restrict its value to a small interval still containing a non-denumerable infinity of alternative values equally consistent with our existence. Furthermore, such an anthropic view either requires that infinitely many alternative universes, many of them with very different values of the fine-structure constant and hence with no observers, actually physically exist, or that some mysterious being singled out our actual universe as the one to be physically instantiated, with an eye to the creation of intelligent beings like ourselves. Is it not in better scientific taste to suppose that the value of the fine-structure constant, or of Planck's constant, will, with the advance of science, prove theoretically calculable? But if this is so (and I have suggested merely one particular way in which it might prove so) at all, then Tomonaga's objection to Schrödinger's original viewpoint must simply one day disappear. But Tomonaga's objection seems to be necessarily wrong because of the non-linearity (theoretically and experimentally required, independently of Schrödinger's view) of the correct rather than the conventional equation. But doesn't this mean *both* that Schrödinger's view can be saved, *and* that a calculation which is to all intents tantamount to a calculation of the value of the fine-structure constant, must be in principle possible? That's the good news; the bad news is that nobody has yet succeeded in doing this latter in a believable manner, and until that is done there is always the possibility that Schrödinger's view does indeed lead to quantization of energy, but not to the empirically correct quantization of energy. But I don't think that Schrödinger would have believed that the deity could be so unkind.

3.3. The problem of locality for a classical relativistic field-theoretical interpretation like Schrödinger's

As we have seen in section 3.2, Schrödinger's objective and realistic interpretation of the quantum theory construes that theory essentially merely as a special sort of classical relativistic field theory. Now the Bell inequalities show that any objective and realistic interpretation of the quantum theory must be non-local in Bell's sense, and it is widely held that non-locality in that sense is incompatible with classical relativistic field theory. Indeed the results of experimental tests of the Bell inequalities were one of the reasons E. T. Jaynes gave for abandoning his own attempts to revive Schrödinger's original programme. Furthermore, the assumption that a classical field theory must be a local theory is crucial to the oft-repeated argument that the unquantized Dirac matter–wave field cannot really be construed as classical field, since the non-commutation of the corresponding field operators at space-like separations entails that the ψ-field is non-local.

I shall maintain that such a view of the relation between locality and classical relativistic field theory is fundamentally mistaken. I argue that, far from requiring locality, classical relativistic field theories, provided they are fully time-symmetric, are inherently non-local theories. Hence, the experimental evidence for non-locality, far from counting against an interpretation of the quantum theory as a classical relativistic field theory, counts (given the abundant independent evidence for time-symmetry in this domain of physics) in favour of such an interpretation of the quantum theory.

Within a relativistic context time-symmetry and non-locality go hand-in-hand for the following reason. Not only is it the case that any non-local causal effect, in the sense of one propagating with a velocity greater than that of light in a given reference frame, will be construed as propagating backwards in time from the point of view of certain other reference frames (this is a trivial consequence of the Lorentz transformation equations for velocities), but it is also the case that in any fully time-symmetric relativistic theory, to any physical causal effects propagating in the ordinary forwards direction of time, there must be physical causal effects propagating backwards in time; i.e. advanced and retarded solutions should in a fully time-symmetric theory be treatable as on a par with one another. However, this means that even if such advanced and retarded causal effects are restricted to the backwards and forwards light cones, combinations of advanced and retarded causal chains can always in principle produce resultants which will appear as non-local causal influences propagating outside the light cones. Actually, even without our assuming that such resultants need occur, it is a feature of Bell's own technical definition of locality that it excludes backwards causation by fiat, by excluding all causal effects propagating outside the forward light cone, and *a fortiori* all merely propagating backwards in time on the backwards light cone. Hence it excludes any fully time-symmetric theory which is not rewritable as an ordinary retarded theory.

An example of such a fully time-symmetric relativistic field theory is classical time-symmetric electrodynamics. It is a widespread but mistaken belief that this theory is rewritable as an ordinary retarded theory and hence a local theory in the sense of Bell.

First I remark that the mere fact that one could always rewrite outgoing advanced waves propagating backwards into the past as incoming retarded waves does not turn the theory into an ordinary retarded theory or render it local in the sense of Bell. In fact it turns it into a very funny kind of retarded theory which is definitely non-local in the sense of Bell. For in such a theory, my choosing at some time in the future to, say, make a charge oscillate, will determine the existence already in the remote past, of incoming retarded radiation which would not be there now were my, as yet unmade, future choice to be different. Moreover, suppose that, in a typical experimental test of a Bell inequality, my choosing to rotate my polarizer were allowed to be a necessary and sufficient condition for the existence of such a retarded incoming wave from the past. Then this wave would pass through the calcium atom and by ordinary scattering by that atom produce a further retarded wave reaching Jones's polarizer and capable of affecting it at a moment which would be at a space-like separation from the event constituting my action. So it would be an easy matter to violate the Bell inequalities with such retarded waves, and they demonstrably fail to satisfy Bell's criterion of locality. You cannot change the physics of a theory by a mere verbal manoeuvre.

Nor is it the case, as Einstein among others seems at one time to have thought (in an early public correspondence with Ritz, but he must have changed his mind later, otherwise he could never have been enthusiastic over Tetrode (1922) and we know he was), that the choice between advanced and retarded solutions or combinations of these depends merely on whether one happens to know the boundary conditions in the past or in the future or a combination of these. Wheeler and Feynman showed definitively that in the absence of their complete absorption condition all sorts of actual physical scenarios occurred in the time-symmetric theory which could not possibly occur in the time-asymmetric theory, indeed ones patently involving backwards causal and non-local effects by Bell's criterion.

Finally I remark that there is no evidence that our actual universe satisfies any boundary condition such as complete absorption which would make time-symmetric electrodynamics equivalent to ordinary retarded electrodynamics, and hence local in the sense of Bell. Most astronomers and cosmologists believe that no empirically and theoretically plausible boundary condition of this kind on our actual universe can be constructed. And in any case the only reason for seeking such a condition is to recover ordinary classical retarded electrodynamics together with the Lorentz–Dirac radiation damping force which Einstein (1917) showed to be empirically false. (He showed that thermal equilibrium between atoms and radiation in a cavity, given the empirical equilibrium distributions of energies, required directed radiative reactions of atoms, i.e. ones incompatible with the Lorentz–Dirac force, and hence with the Wheeler–Feynman (1945) complete absorption condition, but just what one would expect on time-symmetric electrodynamics assuming very incomplete absorption.) There is every reason to suppose that in our actual universe such a time-symmetric classical relativistic theory would retain backwards causality in an essential way and hence be non-local.

The bulk of physicists seem to have believed that any real, rather than merely apparent, element of backwards causation in such a theory must inevitably lead to contradictions, such as the possibility of an experimenter setting off a backwards causal chain which would trigger his own death at an earlier age. But underlying such reasoning there appears to be an elementary error of logic. To prove the consistency of a system of equations in the absence of a certain restrictive condition, it is enough to exhibit one or more physical scenarios which would satisfy the equations in the absence of that restrictive condition. Wheeler and Feynman exhibited several such scenarios in the absence of their complete absorption condition. Hence their system of equations was consistent. But a consistent system of equations simply cannot have any inconsistent solutions. Hence all the existence of causal loops in such a theory could possibly do is, not to generate contradictions but rather generate automatic dynamical constraints on the possible Cauchy data which can be realized on a space-like hypersurface. These will necessarily appear as constraints on the experimental manipulability of initial conditions, that is to say constraints on state preparation. In other words, precisely because the undesirable causal paradoxes are actual contradictions, and because these genuinely would appear if we allowed free construction of Cauchy data, any such theory must automatically generate constraints analogous to the quantum mechanical uncertainty relations. These constraints will also, if one accepts the plausible thesis that macroscopic manipulability and amplification of backwards causal chains would inevitably lead to actual causal paradoxes, automatically ensure that even although such a theory violates micro-causality, it will yield only macroscopic causality in the macroscopic limit. The backwards causation paradoxes thus cannot in principle generate nasty properties in such a fully time-symmetric classical relativistic theory, they instead simply enable us to give very short proofs of the qualitative presence of very nice properties.

That the experimental tests of the Bell inequalities yield non-locality merely as an unproblematic consequence of both relativity and time-symmetry, can be seen directly by examining the time-inverse of such an experiment. We have two photon sources trained on a calcium atom and we are interested in the cases where the upwards transition involving zero angular momentum uptake in the atom occurs. Of course these two sources interfere with one another, and changing the polarization of one will vary the electromagnetic field at the atom and hence vary the probability of the upwards transition in question being produced by those two particular photons. But we add plenty of other sources to the experimental set up, in such a way as to ensure that changing the polarization of one source does not affect the overall probability per unit time that such a transition will occur with the absorption of some pair of photons or other: the set up is such that the experimental manipulation does indeed affect the detailed time development of the electromagnetic field at the location of the atom, but does not affect its statistical properties. Then this unproblematic set-up, in which obviously only forwards causality is at work in ensuring the correlations between source events which trigger transitions, being simply the time-inverse of a test of the Bell inequality, will satisfy the same mathematical conditions. We will not infer any mysterious action at a distance here. We can see how ordinary forwards causality is mediating the correlations via constructive and destructive interference at the atom. Ordinary backwards causality is doing the same thing in the time-inverse set-up. While the condition that the actual overall rate of transitions cannot be experimentally manipulated may seem possible but contrived in the set-up involving only forwards causality, that it will be rather trivially and necessarily fulfilled in the time-inversed set-up already follows from the considerations of the preceding paragraph. This is not a violation of time-symmetry in the dynamics of the theory, but rather follows from the fact that in order to avoid the paradoxes of experimental amplifiability of backwards causal chains, the mathematics of a time-symmetric theory automatically generates a macroscopic causal arrow in the same direction as the cosmologically determined thermodynamic arrow which governs the irreversible actions of experimenters.

We see that we need not in these experiments even postulate any physical effects going backwards and then forwards in time with a resultant outside the light cone. Thus it is sufficient to time-symmetrize Bell's locality condition so that locality only excludes causal effects outside both light cones, to restore locality in a more generalized sense. However, as I have argued, there is no reason for excluding more radically non-local effects from such a theory, for these can easily arise from causal loops. The non-commutation of the fermion field operators at space-like distances could easily be the product of a fully non-local effect generated by a fully-time symmetric classical relativistic theory. I would have expected the correct version of a system of coupled classical relativistic wave equations to be time-symmetric, and I would thus have expected some non-locality to emerge, as it evidently does. Indeed some very mysterious cosmological condition would be necessary in a time-symmetric relativistic theory in order to prevent non-locality arising. There is no way these phenomena can count against Schrödinger's classical relativistic field-theoretic viewpoint.

3.4. Probabilities and wave-packet collapse

Schrödinger's interpretation of the quantum theory was non-probabilistic and possessed no obvious mechanism for wave-packet collapse. One way in which we could resolve both these difficulties simultaneously would be merely to combine Schrödinger's classical field-theoretic picture, which seems to rule out second-quantization, with the Everett–Wheeler many-worlds interpretation of the quantum theory, which can be applied equally to a second-quantized or non-second-quantized version of quantum theory, and is also non-probabilistic and without wave-packet collapse. However, the Everett–Wheeler theory, even if it be coherent, is so riddled with time-asymmetry in its existing versions that it would be hard to reconcile with my earlier sections. So I shall suggest a more time-symmetric and less bizarre solution to the problems of this section. This and the previous section owe something to Cramer (1980) though he is not evidently defending a Schrödinger-style interpretation.

Consider, then, the time-symmetric account of a world containing an emitter atom with its oscillating Schrödinger charge cloud and a potential absorbing atom. If the absorber were absent the emitter would continue to oscillate, emitting negative energy into the past through its advanced radiation and an equal amount of positive energy into the future through its retarded radiation, and hence suffering no overall gain or loss of energy. The relative contributions of the components of the superposition in which the emitter must be in order to produce an overall oscillation of its centre of charge, and thus in order to radiate, do not change with time. There is no exponential decay of one component, contrary to what happens in the retarded theory of Barut and Kraus, because such effects are cancelled out by advanced self-fields. However, if a potential absorber is present, it begins under the influence of the impinging radiation to acquire a charge-cloud oscillation at the same frequency as the emitter. Consider the region in which the absorber is affected by the retarded radiation of the emitter. As the absorber starts to oscillate it emits advanced and retarded radiation, the advanced radiation reacting on the emitter oscillations producing the retarded radiation affecting the absorber. It can be shown, by applications of the usual theory of absorption and its time-inverse, and translating from superpositions of states into charge-cloud oscillations, that if we begin with an emitter in its excited state, with a very small contribution of its ground state due to some slight perturbation, and with the absorber in its ground state, then there will be a phase of mutual enhancement of each other's charge-cloud oscillations followed by a mutual damping of each other's charge-cloud oscillations, which will leave the emitter finally in its ground state and the absorber finally in its excited state.

The energy–momentum transfer occurs in detail as follows. The retarded radiation from the emitter is in phase with the advanced radiation from the absorber along the line joining emitter and absorber, but along this line and beyond the absorber it is out of phase with the retarded radiation from the absorber, and in this region destructive interference occurs. So, along this line the absorber is actually gaining more energy–momentum from the radiation field than it is losing, and the momentum transfer is from the direction of the emitter. Along all other directions the absorber gains precisely as much energy–momentum as it loses. The time-inverse of this situation occurs in the emitter, actual overall energy–momentum loss occurring only, as a result, in the direction of the absorber. All actual energy, momentum and angular momentum lost by the emitter is gained by the absorber. This is not a fulfilment of the Wheeler–Feynman complete absorption conditions which refers rather to all the radiation which would classically have been emitted according to the ordinary retarded theory.

Of course the situation is more complicated if several potential absorbers are present. It may be easier to analyze this situation by considering its time-inverse, where there are several potential emitters and only one potential absorber. We can imagine that there are initially minute charge-cloud fluctuations already present in all potential emitters and in the potential absorber, as a result of some small random background field. The resulting fluctuations in the retarded field at the location of the absorber will at some state or other bring its charge oscillations sufficiently in phase with those momentarily present in one of the emitters to trigger off a mutual amplification and damping sequence. This must depend on a correlation having occurred by chance which is sufficiently large compared with the electromagnetic noise generated by the small random background field. The mutual coupling required seems generally to require that only one emitter can actually effect an energy–momentum transfer to the absorber. But exceptions can occur, such as where all emitters lie in the same straight line with the absorber and the frequency of absorption is the combined frequency of emission (the time-inverse of a gamma-ray track) or, of course, in the time-inverse of an allowed two-photon decay. Hence we see that (with the noted exceptions) in the case where there are several potential absorbers and only one potential emitter, only one will in general couple with the emitter, that this will absorb all the energy, momentum and angular momentum actually lost by the emitter, and that the retarded (and advanced) radiation emitted by the emitter in other directions cannot be absorbed by other absorbers. Here 'cannot' simply means that such absorption would produce advanced fields interfering with the initial emission process and thus preventing its occurrence or causing it to occur in some other way.

If we consider a line through the emitter in some direction other than that of actual absorber, we can see that the radiation along this line can equally be viewed as an incoming and outgoing retarded wave which has merely undergone a phase shift at the location of the emitter. It is thus conceivable that such radiation could prove indistinguishable from a sort of vacuum field or cosmological background field.

It might at first be anticipated that (as with coupled pendula) any transaction processes would immediately repeat itself in reverse, but with advanced waves now playing the role of retarded waves, in a second cycle. But this is not so because the momentum transfer between the two atoms in the exchange produces an effective Doppler red-shift of the radiation, so that it will fall below the resonance frequency. Just one 'photon' will have been more or less irreversibly transferred.

This model appears to reproduce, but now by necessity rather than by fiat, the same general picture as in the programmatic theories of Einstein, Lorentz and de Broglie of the early 1920s: the radiation propagates in an infinity of directions like waves, but the energy–momentum transfer which it mediates propagates like a particle. The resulting analyses of the two-slit experiment, and so on, can be found in Cramer (1980).

Within such a theoretical framework, probabilities will only enter as a result of our ignorance of the precise states of potential absorbers, on the future light cone of the potential emitter, and particularly as a result of our ignorance of the precise phase relations between these states and that of the emitter. One might therefore expect ordinary fully classical probability functions to be generated over the space of such 'hidden variables'. However, we have to remember that phase relations and interference will also be playing a role. In particular we must expect transition probabilities, like quantum-mechanical transition probabilities to exhibit 'interference'. For within the suggested theoretical framework we expect the probability of an actual transition to be proportional both to the amplitude of some retarded field from the emitter at the site of the potential absorber, and proportional to the amplitude of the corresponding advanced field of the absorber at the site of the emitter. But these two amplitudes are symmetrically causally related and simply proportional to one another. Hence, the probability of the transition must be directly proportional to the square of the amplitudes of the retarded field produced by the emitter. Hence to obtain transition probabilities between intial and final states via the transition probabilities associated with hypothetical transitions to intermediate states, we must add amplitude and then square, rather than directly adding probabilities.

This account of radiation wave-packet 'collapse' and of the emergence of quantum probabilities needs of course, as Cramer emphasized, to be generalized to the context of a fully time-symmetric account of the matter–wave fields, if it is also to explain the collapse of matter–wave packets. Presumably one will find here that not only actual transfer of momentum, energy and angular momentum, but also actual transfer of mass and charge, can only occur in a fully time-symmetric theory along the lines joining emitters and actual absorbers. So the suggestion of this section is that, though we are far from satisfactorily solving some of the mathematical problems involved in this sort of time-symmetric theory, it may be the case that we do not after all need to go outside the context of classical relativistic field theories, in order to solve the remaining quantum mysteries.

Conclusion

Schrödinger's approach to physics was characterized by the fact that he was not prepared to accept the emasculated standards of physical explanation so warmly embraced by some of his contemporaries and by almost all his successors. But it seems to me largely a sociological accident that subsequent 20th-century physics abandoned Schrödinger's standards and has gone the way it has. Schrödinger's original programme for the quantum theory as an essentially classical field theory did not *have* to be abandoned. It merely ran into difficulties. But, as Paul Feyerabend has stressed, it only takes a little imagination and a little rhetoric to construct a plausible defence of almost any ambitious theoretical programme, whatever the *prima facie* objections to it. While my rhetorical defence of Schrödinger's original programme relies heavily on the apparent properties of time-symmetric classical relativistic theories, such a defence is not in fact anachronistic, for Schrödinger was already familiar with Tetrode's 1922 formulation of time-symmetric electrodynamics and with Tetrode's well-argued suggestion in that paper (which also impressed Einstein) that this theory possessed precisely the properties needed to solve the quantum riddle. It seems to me a mere historical accident that such a theory was not early on combined with Schrödinger's classical interpretation of wave-mechanics, in its relativistic version, and that such a combination did not become the conventional orthodoxy. Of course such a combination still poses some unsolved problems, but they do not seem to me any worse than the problems which afflict the much stranger view which has become the conventional orthodoxy. Given Feynman's, Jaynes's and Barut and Kraus's discoveries that merely first-quantized theory does far more than one might reasonably have expected of it, it does not seem to me that the more conventional approach will much longer retain any real advantage in terms of confirmed quantitative predictions. In fact it seems to me, though I am scarcely an expert, now an entirely open theoretical possibility that conventionally interpreted second-quantized theory, in so far as that theory is consistent at all, may eventually turn out, as far as its predictive content is concerned, to be merely a conservative extension of a consistently formulated and classically interpreted fully time-symmetric first-quantized theory. Be this as it may, as a historian of physics with a somewhat wry perspective on 20th-century physics, it seems to me that we do not know (or at any rate I do not know) enough, yet, to state with any conviction that Schrödinger's original interpretation of his equation was merely a blind alley, a wrong turning, or to state that his and Einstein's quixotic refusal to abandon classical standards of physical explanation was the action of heretics and sinners rather than of not yet canonized saints and martyrs. My own guess is that any theory of the physical universe based on the mathematics of the continuum, be it Schrödinger's essentially classical field theory, or orthodox second-quantized field theory, is ultimately doomed to failure and paradox. But only the future history of physics, and certainly not the present state of play, can settle such issues.

References

d'Abro, A. (1939, 1951) *The Rise of the New Physics* (original title *Decline of Mechanism*), chap. 33. van Nostrand 1939, Dover 1951
Barut, A. O. (1978,) *Z. Naturforsch.* 33a, 993–4
Barut, A. O. and Kraus, J. (1983) *Foundations of Physics* 13, 189–94
Cramer, J. G. (1980) *Phys. Rev. D*, 22, 362–78
Einstein, A. (1917) *Physik. Zeits.* 18, 121
Feynman, R. P. (1950) *Phys. Rev.* 80, 440–57
Jaynes, E. T. (1973) in *Coherence and Quantum Optics* (Mandel, L. and Wolf, E., eds.), pp. 35–81. Plenum Press
Samec, A. (1980) *Hadronic J.* 3, 832–53
Schiff, L. I. (1955) *Quantum Mechanics*, 2nd edn., pp. 209–13. McGraw-Hill
Tetrode, H. (1922) *Zeits f. Physik* 10, 317
Tomonaga, S.-I. (1966) *Quantum Mechanics*, vol. 2, chap. 6. North Holland
Wheeler, J. A. and Feynman, R. P. (1945) *Rev. Mod. Phys.* 17, 157–81
Wyler, A. (1969) *C. R. Acad. Sci. Paris* 269 A, 743

4

Are there quantum jumps?

J.S. BELL

CERN-TH, Switzerland

If we have to go on with these damned quantum jumps, then I'm sorry that I ever got involved.

E. Schrödinger

4.1. Introduction

I have borrowed the title of a characteristic paper by Schrödinger (Schrödinger, 1952). In it he contrasts the smooth evolution of the Schrödinger wavefunction with the erratic behaviour of the picture by which the wavefunction is usually supplemented, or 'interpreted', in the minds of most physicists. He objects in particular to the notion of 'stationary states', and above all to 'quantum jumping' between those states. He regards these concepts as hangovers from the old Bohr quantum theory, of 1913, and entirely unmotivated by anything in the mathematics of the new theory of 1926. He would like to regard the wavefunction itself as the complete picture, and completely determined by the Schrödinger equation, and so evolving smoothly without 'quantum jumps'. Nor would he have 'particles' in the picture. At an early stage, he had tried to replace 'particles' by wavepackets (Schrödinger, 1926). But wavepackets diffuse. And the paper of 1952 ends, rather lamely, with the admission that Schrödinger does not see how, for the present, to account for particle tracks in track chambers ... nor, more generally, for the definiteness, the particularity, of the world of experience, as compared with the indefiniteness, the waviness, of the wavefunction. It is the problem that he had had (Schrödinger, 1935*a*) with his cat. He thought that she could not be both dead and alive. But the wavefunction showed no such commitment, superposing the possibilities. Either the wavefunction, as given by the Schrödinger equation, is not everything, or it is not right.

Of these two possibilities, that the wavefunction is not everything, or not right, the first is developed especially in the de Broglie–Bohm 'pilot wave' picture. Absurdly, such theories are known as 'hidden variable' theories. Absurdly, for there it is not in the wavefunction that one finds an image of the visible world, and the results of experiments, but in the complementary 'hidden'(!) variables. Of course the extra variables are not confined to the visible 'macroscopic' scale. For no sharp definition of such a scale could be made. The 'microscopic' aspect of the complementary variable is indeed hidden from us. But to admit things not visible to the gross creatures that we are is, in my opinion, to show a decent humility, and not just a lamentable addiction to metaphysics. In any case, the most hidden of all variables, in the pilot wave picture, is the wavefunction, which manifests itself to us only by its influence on the complementary variables.

If, with Schrödinger, we reject extra variables, then we must allow that his equation is not always right. I do not know that he contemplated this conclusion, but it seems to me inescapable. Anyway it is the line that I will follow here. The idea of a small change in the mathematics of the wavefunction, one that would little affect small systems, but would become important in large systems, like cats and other scientific instruments, has often been entertained. It seems to me that a recent idea (Ghirardi, Rimini and Weber, 1985), a specific form of spontaneous wavefunction collapse, is particularly simple and effective. I will present it below. Then I will consider what light it throws on another of Schrödinger's preoccupations. He was one of those who reacted most vigorously (Schrödinger, 1935a, b, 1936) to the famous paper of Einstein, Podolsky and Rosen (1935). As regards what he called 'quantum entanglement', and the resulting EPR correlations, he 'would not call that *one* but rather *the* characteristic trait of quantum mechanics, the one that enforces its entire departure from classical lines of thought'.

4.2. Ghirardi, Rimini and Weber

The proposal of Ghirardi, Rimini and Weber, is formulated for nonrelativistic Schrödinger quantum mechanics. The idea is that while a wavefunction

$$\psi(t, \mathbf{r}_1, \mathbf{r}_2, \dots, \mathbf{r}_N) \qquad (2.1)$$

normally evolves according to the Schrödinger equation, from time to time it makes a jump. Yes, a jump! But we will see that these GRW jumps have little to do with those to which Schrödinger objected so strongly. The only resemblance is that they are random and spontaneous. The probability per unit time for a GRW jump is

$$\frac{N}{\tau}, \tag{2.2}$$

where N is the number of arguments $\mathbf{r}$ in the wavefunction, and τ is a new constant of nature. The jump is to a 'reduced' or 'collapsed' wavefunction

$$\psi' = \frac{j(\mathbf{x} - \mathbf{r}_n)\psi(t, \ldots)}{R_n(\mathbf{x})}, \tag{2.3}$$

where $\mathbf{r}_n$ is randomly chosen from the arguments $\mathbf{r}$. The jump factor j is normalized:

$$\int d^3\mathbf{x} \, |j(\mathbf{x})|^2 = 1. \tag{2.4}$$

Ghirardi, Rimini and Weber suggest a Gaussian:

$$j(\mathbf{x}) = K \exp(-\mathbf{x}^2/2a^2) \tag{2.5}$$

where a is again a new constant of nature. R is a renormalization factor:

$$|R_n(\mathbf{x})|^2 = \int d^3\mathbf{r}_1 \ldots d^3\mathbf{e}_N \, |j\psi|^2. \tag{2.6}$$

Finally the collapse centre $\mathbf{x}$ is randomly chosen with probability distribution

$$d^3\mathbf{x} \, |R_n(\mathbf{x})|^2. \tag{2.7}$$

For the new constants of nature, GRW suggest as orders of magnitude

$$\tau \approx 10^{15} \, \text{s} \approx 10^8 \, \text{year} \tag{2.8}$$

$$a \approx 10^{-5} \, \text{cm}. \tag{2.9}$$

An immediate objection to the GRW spontaneous wavefunction collapse is that it does not respect the symmetry or antisymmetry required for 'identical particles'. But this will be taken care of when the idea is developed in the field theory context, with the GRW reduction applied to 'field variables' rather than 'particle positions'. I do not see why that should not be possible, although novel renormalization problems may arise.

There is no problem in dealing with 'spin'. The wavefunctions ψ and ψ' in (2.3) can be supposed to carry suppressed spin indices.

Consider now the wavefunction

$$\phi(\mathbf{s}_1 \ldots \mathbf{s}_L)\chi(\mathbf{r}_1 \ldots \mathbf{r}_M), \tag{2.10}$$

where L is not very big and M is very very big. The first factor, ϕ, might represent a small system, for example an atom or molecule, that is temporarily isolated from the rest of the world ... the latter, or part of it, represented by the second factor, χ. The GRW process for the complete wavefunction implies independent GRW processes for the two factors. From (2.8) we can forget about GRW processes in the small system. But in the big system, with M of order say 10^{20} or larger, the mean lifetime before a GRW jump is some

$$\frac{10^{15}}{10^{20}} = 10^{-5} \text{ s} \tag{2.11}$$

or less.

Consider next a wavefunction like

$$\phi_1(\mathbf{s}_1 \ldots \mathbf{s}_L)\chi_1(\mathbf{r}_1 \ldots \mathbf{r}_M) + \phi_2(\mathbf{s}_1 \ldots \mathbf{s}_L)\chi_2(\mathbf{r}_1 \ldots \mathbf{r}_M). \tag{2.12}$$

This might represent the aftermath of a 'quantum measurement' situation. Some 'property' of the small system has been 'measured' by interaction with a large 'instrument', which is thrown as a result into one or other of the states χ_1 or χ_2, corresponding to different pointer readings. This macroscopic difference between χ_1 and χ_2 implies that, for very many arguments $\mathbf{r}$, multiplication of the wavefunction by $j(\mathbf{x} - \mathbf{r})$ will reduce to zero one or other of the terms in (2.12). Thus in a time of order (2.11) one of the terms will disappear, and only the other will propagate. The wavefunction commits itself very quickly to one pointer reading or the other. Moreover, the probability that one term rather than the other survives is proportional to the fraction of the total norm which it carries – in agreement with the rule of pragmatic quantum theory.

Quite generally any embarrassing macroscopic ambiguity in the usual theory is only momentary in the GRW theory. The cat is not both dead and alive for more than a split second. One could worry perhaps if the GRW process does not go too far. In the usual pragmatic theory the 'reduction' or 'collapse' of the wavefunction is an operation performed by the theorist at some convenient time. Usually it will be delayed till the Schrödinger equation has established a very big difference between χ_1 and χ_2. The GRW process is one of nature, and comes about as soon as the difference between χ_1 and χ_2 is big enough. I think that with suitable values of the natural constants (2.8, 2.9) the GRW theory will nevertheless agree with the pragmatic theory in practice. But studies on models would be useful to build up confidence in this.

4.3. Quantum entanglement

There is nothing in this theory but the wavefunction. It is in the wavefunction that we must find an image of the physical world, and in particular of the arrangement of things in ordinary 3-dimensional space. But the wavefunction as a whole lives in a much bigger space, of 3N-dimensions. It makes no sense to ask for the amplitude or phase or whatever of the wavefunction at a point in ordinary space. It has neither amplitude nor phase nor anything else until a multitude of points in ordinary 3-space are specified. However, the GRW jumps (which are part of the wavefunction, not something else) are well localized in ordinary space. Indeed each is centred on a particular spacetime point $(\mathbf{x}, t)$. So we can propose these events as the basis of the 'local beables' of the theory. These are the mathematical counterparts in the theory to real events at definite places and times in the real world (as distinct from the many purely mathematical constructions that occur in the working out of physical theories, as distinct from things which may be real but not localized, and as distinct from the 'observables' of other formulations of quantum mechanics, for which we have no use here). A piece of matter then is a galaxy of such events. As a schematic psychophysical parallelism we can suppose that our personal experience is more or less directly of events in particular pieces of matter, our brains, which events are in turn correlated with events in our bodies as a whole, and they in turn with events in the outer world.

In this paper we will use the notion of localization of events only in a rough way. We will localize them in one or other of two widely separated regions of space which we suppose to be occupied by two widely separated systems.

Let the arguments $\mathbf{s}$ and $\mathbf{r}$ in (2.12) refer to the two sides, respectively, in an Einstein–Podolsky–Rosen–Bohm setup, with L as well as M now large. A source, which for simplicity we omit from the analysis, emits a pair of spin $\frac{1}{2}$ neutrons in the singlet spin state. They move through Stern–Gerlach magnets to counters which register for each neutron whether it has been deflected 'up' or 'down' in the corresponding magnet. According to the Schrödinger equation the wavefunction would come out like (2.12), with ϕ_1 or ϕ_2 corresponding to 'up' or 'down' on the left, and χ_1 or χ_2 corresponding to 'down' and 'up' on the right. Suppose that the left hand counters are closer to the source, and so register before the right hand ones. That is to say, suppose that ϕ_1 differs macroscopically from ϕ_2 before χ_1 from χ_2. Then the GRW jumps on the left quickly reduce the wavefunction to one or other of the two terms in (2.12). The choice between χ_1 and χ_2, as well as between ϕ_1 and ϕ_2, has then been made. The jumps on the left are decisive, and those on the right have no opportunity to be so.

In all this the GRW account is very close to that of a common way of presenting conventional quantum mechanics, with 'measurement' causing 'wavefunction collapse' – and with a 'measurement' somewhere causing 'collapse' everywhere. But it is important that in the GRW theory everything, including 'measurement', goes according to the mathematical equations of the theory. Those equations are not disregarded from time to time on the basis of supplementary, imprecise, verbal, prescriptions.

In this EPRB situation, an 'up' on the left implies a subsequent 'down' on the right, and vice versa. Now of course it was not the existence of correlations between distant events that scandalized EPR, and led Einstein (Einstein, 1949) to use the word 'paradox' in this connection. Such correlations are common in daily life. If I find that I have brought only one glove, the left handed, then I confidently predict that the one at home will be found to be right handed. In the everyday conception of things there is no puzzle here. Both gloves have been there all morning, and each has been right or left handed all the time. Observation of the one taken from my pocket gives information about, but does not influence, the one left at home. As regards EPRB correlations, what is disturbing about quantum mechanics, especially as sharpened by GRW, is that before the first 'measurement' there *is* nothing but the quantum mechanical wavefunction – entirely neutral between the two possibilities. The decision between these possibilities is made for both of the mutually distant systems only by the first 'measurement' on one of them. There is no question, if there *was* nothing but the wavefunction, of just revealing a decision already taken. It was this 'spooky action at a distance', the immediate determining of events in a distant system by events in a near system, that scandalized EPR. They concluded that quantum mechanics must, at best, be incomplete. There must be in nature additional variables, not yet known to quantum mechanics, in both systems, which determine in advance the results of experiments, and which happen to have become correlated at the source – just as gloves happen to be sold in matching pairs.

It is now very difficult to maintain this hope, that local causality might be restored to quantum mechanics by the addition of complementary variables. The perfect correlations actually considered by EPR, with parallel polarizers in the EPRB setup, do not present any difficulty in this respect. But the imperfect correlations implied by quantum mechanics, for misaligned polarizers, prove more intractable (e.g. Bell, 1981).

The GRW theory does not add variables. But by adding mathematical precision to the jumps in the wavefunction, it seems simple to make precise the action at a distance of ordinary quantum mechanics. The most disturbing aspect of this is the apparent difficulty of reconciling it with Lorentz invariance. For in a Lorentz invariant theory we tend to think that 'nothing goes faster than light'. So we turn now to a discussion of Lorentz invariance.

4.4. Relative time translation invariance

Of course we cannot discuss full Lorentz invariance in the context of the nonrelativistic model presented above. But there is a residue, or at least an analogue, of Lorentz invariance, which can be discussed in the case of two widely separated systems. Consider the Lorentz transformation

$$z' = \gamma(z - vt), \qquad t' = \gamma(t - vz) \tag{4.1}$$

with x and y unchanged, where the velocity of light has been set equal to unity, and

$$\gamma = \frac{1}{(1 - v^2)^{1/2}}. \tag{4.2}$$

In the case of a system at a large distance, a, from the origin, it is convenient to introduce a new origin, so that

$$z \rightarrow z + a. \tag{4.3}$$

Then (4.1) becomes

$$z' = -a + \gamma(z + a - vt), \qquad t' = \gamma(t - v(z + a)). \tag{4.4}$$

Taking v very small and a very large so that

$$va = k \tag{4.5}$$

(4.4) becomes

$$z' = z, \qquad t' = t - k. \tag{4.6}$$

In the case of a single system this tells us simply to expect invariance with respect to translation in time. But in the case of two systems displaced from the origin in opposite directions, and so with different signs for k, it tells us to expect invariance with respect to displacement in *relative* time.

Multiple time formalism, with independent times for different particles, or for different points in space, is an old story in relativistic quantum theory. It is less familiar in the context of the nonrelativistic theory. However, it is easily implemented in the case of *noninteracting systems* at the level of the Schrödinger equation. Let two noninteracting subsystems have separate Hamiltonians A and B, respectively, so that the total Hamiltonian is

$$H = A + B. \tag{4.7}$$

Then from the ordinary 1-time wavefunction $\psi(t, \ldots)$ we can define a 2-

time wavefunction

$$\psi(t', t'', \ldots) = \exp \frac{i(t-t')A}{\hbar} \exp \frac{i(t-t'')B}{\hbar} \psi(t, \ldots). \qquad (4.8)$$

Since A and B commute, the relative order of the two exponentials in (4.8) is unimportant. (However, if A and B are time-dependent, the two exponentials must separately be time ordered, as in (A.5).) The 2-time wavefunction satisfies the two Schrödinger equations

$$\hbar i \frac{\partial}{\partial t'} \psi(t', t'' \ldots) = A\psi(t', t'', \ldots) \qquad (4.9)$$

$$\hbar i \frac{\partial}{\partial t''} \psi(t', t'' \ldots) = B\psi(t', t'', \ldots). \qquad (4.10)$$

These equations are invariant against independent shifts in the origins of the two time variables (provided any time dependent external fields in A and B are shifted appropriately).

It remains to see if this relative time variance survives the introduction of the GRW jumps. It does. I did not find a short elegant argument, and have relegated the clumsy arguments that I did find to an appendix. From the ordinary 1-time wavefunction for time i, a 2-time wavefunction can again be constructed. It incorporates the jumps of subsystem-1 between times i and i', and those of subsystem-2 between i and i''. In terms of this a formula can be found (A.22, A.23) for the probability of subsequent jumps before times f' and f'' in the two subsystems respectively. It can be interpreted as supplementing (4.9, 4.10) by giving the probabilities for jumps in the two systems as t' and t'' are advanced independently from independent starting points. It does not depend on t' or t'' except through the 2-time wavefunction ψ (and any time dependent external fields in Hamiltonians A and B). The relative time translation invariance of the theory is then manifest.

The reformulation (A.22, A.23) of the theory can also be used to calculate the statistics of jumps in one system separately, disregarding what happens in the other. The result (A.24, A.25) makes no reference to the second system. Events in one system, considered separately, allow no inference about events in the other, nor about external fields at work in the other, nor even about the very existence of the other system. There are no 'messages' in one system from the other. The inexplicable correlations of quantum mechanics do not give rise to signalling between noninteracting systems. Of course, however, there may be correlations (e.g. those of EPRB) and if something about the second system is given (e.g. that it is the other side of an EPRB setup) and something about the overall state (e.g. that it is the

EPRB singlet state) then inferences from events in one system (e.g. 'yes' from the 'up' counter) to events in the other (e.g. 'yes' from the 'down' counter) are possible.

4.5. Conclusion

I think that Schrödinger could hardly have found very compelling the GRW theory as expounded here – with the arbitrariness of the jump function, and the elusiveness of the new physical constants. But he might have seen in it a hint of something good to come. He would have liked, I think, that the theory is completely determined by the equations, which do not have to be talked away from time to time. He would have liked the complete absence of particles from the theory, and yet the emergence of 'particle tracks', and more generally of the 'particularity' of the world, on the macroscopic level. He might not have liked the GRW jumps, but he would have disliked them less than the old quantum jumps of his time. And he would not have been at all disturbed by their indeterminism. For as early as 1922, following his teacher Exner, he was expecting the fundamental laws to be statistical in character: '. . . once we have discarded our rooted predilection for absolute Causality, we shall succeed in overcoming the difficulties . . .' (Schrödinger, 1957).

For myself, I see the GRW model as a very nice illustration of how quantum mechanics, to become rational, requires only a change which is very small (on some measures!). And I am particularly struck by the fact that the model is as Lorentz invariant as it could be in the nonrelativistic version. It takes away the ground of my fear that any exact formulation of quantum mechanics must conflict with fundamental invariance.

Appendix

Let

$$P(f; \mathbf{x}_m, n_m, t_m; \ldots \mathbf{x}_1, n_1, t_1; i) \mathrm{d}^3 x_1 \ldots \mathrm{d}^3 x_m \, \mathrm{d}t_1 \ldots \mathrm{d}t_m \qquad (\text{A.1})$$

be the probability that between some time i and some later time f there are m jumps, with the first at time t_1 in the interval $\mathrm{d}t_1$, involving argument $\mathbf{r}_{n_1}$, and centred at $\mathbf{x}_1$ in $\mathrm{d}^3 x_1$; and with the second at time t_2, involving argument $\mathbf{r}_{n_2}$ centred at $\mathbf{x}_2, \ldots$ and so on. Then, from the basic assumptions,

$$P = \exp \lambda N(i - f) \langle i | E^+(f, i) E(f, i) | i \rangle, \qquad (\text{A.2})$$

where N is the total 'particle number', $|i\rangle$ denotes the initial state

$$|i\rangle = \psi(i, \mathbf{r}_1, \mathbf{r}_2 \ldots) \qquad (\text{A.3})$$

and
$$E(f, i) = U(f, t_m)j(n_m, \mathbf{x}_m) \dots U(t_2, t_1)j(n_1, \mathbf{x}_1)U(t_1, i) \qquad (A.4)$$
with
$$U(s, t) = T \exp \int_s^t dt' \frac{H(t')}{ih} \qquad (A.5)$$
and
$$j(n, x) = \lambda^{1/2}j(\mathbf{x} - \mathbf{r}_n). \qquad (A.6)$$

In (A.5) we allow that the Hamiltonian might be time dependent, and so have a time-ordered product. Note the unitarity relation
$$U^+ U = 1. \qquad (A.7)$$
The leftmost U in (A.4) is actually redundant in (A.2), because of (A.7), but it is convenient later. The exponential in front of (A.2) arises from a product of exponentials
$$\exp - \lambda N(t' - t),$$
which are the probabilities of having no jumps in the corresponding time intervals. The formulae could be simplified somewhat by introducing Heisenberg operators, but we will not do so here.

Let us calculate from (A.1)–(A.4), for given i, the conditional probability distribution for jumps in the interval i' till f when the jumps between i and i' are given. We have only to divide (A.1) by the probability for the given jumps:
$$\exp \lambda N(i - i')|R|^2 d^3\mathbf{x}_1 \dots dt_1 \dots \qquad (A.8)$$
with, from (A.2),
$$|R|^2 = \langle i|E^+(i', i)E(i', i)|i \rangle. \qquad (A.9)$$
The result may be expressed in terms of
$$|i' \rangle = \frac{E(i', i)|i \rangle}{R} \qquad (A.10)$$
when we note the factorization property
$$E(f, i) = E(f, i')E(i', i). \qquad (A.11)$$
If we renumber the jumps in the reduced interval after i' to begin again with 1, we find again just (A.1)–(A.4) with i replaced everywhere by i'. So this was only a rather elaborate consistency check. But the manipulations involved will be useful for another purpose in a moment.

Let us now calculate from (A.1)–(A.4), with fixed f, the probability P' for jumps specified only up to some earlier time f', regardless of what happens later. To do so we have to sum over all possibilities in the interval between f' and f. There might be $0, 1, 2, \dots$ extra jumps in that remaining interval

interval. The probability of the given jumps in the reduced interval, and no jumps in the remainder, is given directly by (A.2), which we rewrite as

$$X_0 \exp \lambda N(i-f') \langle i | E^+(f', i) E(f', i) i \rangle \qquad (A.12)$$

with

$$X_0 = \exp \lambda N(f'-f). \qquad (A.13)$$

With one extra jump, E^+E in the expectation value is replaced by

$$E^+ U^+ |j(n, x)|^2 UE, \qquad (A.14)$$

where the extra factor U evolves the system from time f' till the time t of the extra jump (n, x). Integration over x, using (2.4), replaces $|j(n, x)|^2$ by λ. The extra U^+U then goes away by unitarity. Summation over n gives a factor N, and integration over time t gives a factor $(f-f')$. Then the total one extra jump contribution to P' is (A.12) with X_0 replaced by

$$X_1 = \lambda N(f-f') \exp \lambda N(f'-f). \qquad (A.15)$$

Proceeding in this way we find for the n-extra-jump contribution to P' again (A.11) but with X_0 replaced by

$$X_n = \frac{(\lambda N(f-f'))^n}{n!} \exp \lambda N(f'-f). \qquad (A.16)$$

The factor $n!$ arises from the restriction of the multiple time integral to chronological order. To obtain the total P' we have to sum these n-extra-jump contributions over all n. This is easy, for

$$\sum X_n = 1. \qquad (A.17)$$

The result for P' is just (A.1)–(A.4) with f replaced by f'. This is only as expected, but similar manipulations will be useful below.

Suppose now that the system falls into two noninteracting subsystems, with commuting Hamiltonians A and B, respectively:

$$H = A + B. \qquad (A.18)$$

Then the operators U factorize:

$$U(t', t) = V(t', t) W(t', t) \qquad (A.19)$$

with V and W constructed like U in (A.5), but with A and B replacing H. Since V and W commute, we can collect together the factors referring to each subsystem in (A.2), with the result

$$P = \exp \lambda L(i-f) \exp \lambda M(i-f) \langle i | F^+ F G^+ G | i \rangle, \qquad (A.20)$$

where F and G are constructed like E in (A.4) but with operators of the first and second subsystems, respectively. The integers L and M are the 'particle numbers' of the subsystems:

$$L + M = N. \qquad (A.21)$$

At this stage the initial and final times i and f are common to the two subsystems. But by the manipulations described above we can pass from i to f to later initial times, and earlier final times. Moreover, because the jump and evolution operators commute with one another, and have been collected together into separate commuting factors F and G, this can be done independently for the two subsystems. So we can take independent initial times i' and i'', and independent final times f' and f'', for the two subsystems, respectively.

The resulting probability distribution, over jumps in the reduced time intervals, is

$$P(f', f''; \mathbf{x}_m, n_m, t_m; \ldots x_1, n_1, t_1; i', i'') d^3\mathbf{x}_1 \ldots d^3\mathbf{x}_m \, dt_1 \ldots dt_m, \quad (A.22)$$

where

$$P = \langle i', i'' | F^+ F G^+ G | i', i'' \rangle. \quad (A.23)$$

The jumps and evolutions before i' and i'', in the two subsystems, respectively, have been incorporated into the initial state $|i', i''\rangle$. The jumps and evolutions in the reduced intervals, i' till f' till f'', make F and G, as in (A.4).

Note finally that if we are interested only in what happens in subsystem 1, we can sum over all possibilities for the second system in a now familiar way. The result is just (A.22), with reference to jumps in system 1 only, and (A.23) without any operator G. It is equivalent to

$$P = \text{trace}_1 F^+ F \rho, \quad (A.24)$$

where the trace is over the state space of system 1, and

$$\rho = \text{trace}_2 |i', i''\rangle\langle i', i''| \quad (A.25)$$

with the trace over the state space of system 2.

References

Bell, J. S. (1981) *J. de Physique* 42, c2, 41–61
Einstein, A. (1949) *Reply to criticisms. Albert Einstein, Philosopher and Scientist* (Schilp, P. A., ed.). Tudor
Einstein, A., Podolsky, B. and Rosen, N. (1935) *Phys. Rev.* 47, 777
Ghirardi, G. C., Rimini, A. and Weber, T. (1985) *Phys. Rev. D* 32, 470
Schrödinger, E. (1926) *Annal. Phys.* 79, 489–527
Schrödinger, E. (1935a) *Naturwissenschaften* 23, 807–12, 823–8, 844–9
Schrödinger, E. (1935b) *Proc. Camb. Phil. Soc.* 31, 555–63
Schrödinger, E. (1936) *Proc. Camb. Phil. Soc.* 32, 446–52
Schrödinger, E. (1952) *Brit. J. Phil. Sci.* 3, 109–23, 233–47
Schrödinger, E. (1957) *What is a law of nature? Science Theory and Man*, pp. 133–47

5

Square root of minus one, complex phases and Erwin Schrödinger

CHEN NING YANG
State University of New York

5.1. Introduction

In a lecture in April 1970 Dirac talked about the early days of quantum mechanics (Dirac, 1972). Among other topics he discussed noncommutative algebra, and added

> The question arises whether the noncommutation is really the main new idea of quantum mechanics. Previously I always thought it was but recently I have begun to doubt it and to think that maybe from the physical point of view, the noncommutation is not the only important idea and there is perhaps some deeper idea, some deeper change in our ordinary concepts which is brought about by quantum mechanics.

He then expanded on this subject and concluded

> So if one asks what is the main feature of quantum mechanics, I feel inclined now to say that it is not noncommutative algebra. It is the existence of probability amplitudes which underlie all atomic processes. Now a probability amplitude is related to experiment but only partially. The square of its modulus is something that we can observe. That is the probability which the experimental people get. But besides that there is a phase, a number of modulus unity which can modify without affecting the square of the modulus. And this phase is all important because it is the source of all interference phenomena but its physical significance is obscure. So the real genius of Heisenberg and Schrödinger, you might say, was to discover the existence of probability amplitudes containing this phase quantity which is very well hidden in nature and it is because it was so well hidden that people hadn't thought of quantum mechanics much earlier.

One may or may not agree with Dirac on the question of which was more important: the introduction of an amplitude with a phase or that of noncommutative algebra, but there is no doubt that both are revolutionary developments of profound significance in the physicists' description of nature.

Classical physics, that is the physics before 1925, used exclusively real quantities. This was true for mechanics, thermodynamics, electrodynamics – the whole of classical physics. To be sure, complex numbers were used in many places. For example, in solving a linear alternating current problem complex numbers were used. But after a solution had been found, one always took the real or imaginary part of the solution in order to obtain the *true* physical answer. So the use of complex numbers was as a computational aid, i.e. the physics was conceptually in terms of real numbers.

With matrix mechanics and wave mechanics, however, the situation dramatically changed. Complex numbers became a conceptual element of the very foundation of physics: the fundamental equations of matrix mechanics and of wave mechanics:

$$pq - qp = -i\hbar \tag{1.1}$$

$$i\hbar \frac{\partial \psi}{\partial t} = H\psi \tag{1.2}$$

both explicitly contain the imaginary unit $i = \sqrt{-1}$. It is to be emphasized that the very meaning of these equations would be totally destroyed if one tries to get rid of i by writing (1.1) and (1.2) in terms of real and imaginary parts.

5.2. Complex numbers in matrix and wave mechanics

The following is a brief history of the entry of complex numbers in matrix mechanics and in wave mechanics.

Take matrix mechanics first. In the pioneering paper of Heisenberg (1925) a comparison was made between the Fourier transform of a dynamical quantity (which depends on one state and one Fourier multiplicity) and its 'quantum theoretical' correspondence (which depends upon two states). In this process Heisenberg very naturally was conceptually discussing *complex* Fourier amplitudes. In the subsequent two-man paper (Born and Jordan, 1925), (1.1) explicitly appeared for the first time in history. That was also the first time that the imaginary i entered physics in a fundamental way. A little later, in Dirac's first paper on quantum mechanics (Dirac, 1925), (1.1) again appeared, together with

$$\dot{q} = [q, H] = (qH - Hq)/(i\hbar) \tag{2.1}$$

which also explicitly contains i. These developments implied that complex numbers play an essential role in matrix mechanics. But there seemed to be little appreciation at the time that this was a major new development in physics – perhaps because matrix mechanics was so new and Fourier analysis so natural that the full implications of the entry of complex numbers was obscured by the great revolution that was taking place.

Now we turn to wave mechanics which was created* in a historical series of six papers (Schrödinger, 1926*a–f*) all written within the first six months of 1926 by Erwin Schrödinger. In the first five of these Schrödinger had in mind the factorization of his wave function into a real stationary function of x and a sinusoidal function of time (Schrödinger, 1926*c*)†.

That Schrödinger did this was not surprising, since he was thinking of a standing wave description of the electron, very much in analogy with a standing electromagnetic wave or a water wave. Such waves do have phases, but nevertheless they are described by real functions of space-time. In Schrödinger (1926*e*) for example, there appeared a footnote to the equation

$$\psi_n = e^{-x^2/2} H_n(x) e^{2\pi i \nu_n t} \tag{2.2}$$

which reads 'i means $\sqrt{-1}$. On the right-hand side the real part is to be taken, *as usual*.' [my italics], revealing his general attitude on this matter, which was the same as in the usual *linear* circuit theory: ψ may be complex, but one always takes the real part in the end.

Of course, in his search for the relationship between matrix mechanics and wave mechanics Schrödinger unavoidably encountered $i = \sqrt{-1}$, as for example in equation (20) Schrödinger (1926*c*). Whether this fact perturbed him we shall probably never know. But he must have been disturbed when he involved himself with a discussion of quadratic forms such as $\psi(\partial\bar\psi/\partial t)$ (which he did briefly in Schrödinger (1926*c*)) or $\psi\bar\psi$ (which he did sometime before June 6, 1926, see below).

On May 27, 1926, H. A. Lorentz, then 73 years old, wrote a long letter to Schrödinger thanking the latter for having sent him the proof sheets of three articles and raising a lot of questions, some quite general, others very specific, about wave mechanics. Two of these questions are relevant to our present discussion: (*a*) how to interpret the ψ function for two or more particles; (*b*) Lorentz's opinion that 'the true "equations of motion" . . . [should not] contain E at all, but contain time derivatives instead.' Schrödinger answered on June 6 in an equally long letter of eight points. The first two points addressed the two questions of Lorentz's mentioned above.

About (*a*), Schrödinger said that he had abandoned the expression $\psi(\partial\bar\psi/\partial t)$ of his earlier manuscript (Schrödinger, 1926*f*), and was now

---

* Pais (1986) quotes Weyl as saying 'Schrödinger did his great work during a late erotic outburst in his life'.

† See remark in parentheses following equation (35).

focussing on $\psi\bar{\psi}$ for the electric charge density in real space. He then continued: 'What is unpleasant here, and indeed directly to be objected to, is the use of complex numbers. ψ is surely fundamentally a real function.' There followed an involved suggestion of how to generate a complex ψ from its real part ψ_r, a suggestion clearly not quite satisfactory to Schrödinger himself.

About (*b*), Schrödinger wrote down

$$-\hbar^2\ddot{\psi} = E^2\psi \tag{2.3}$$

and then eliminated E by using $H\psi = E\psi$ to obtain

$$-\hbar^2\ddot{\psi} = H^2\psi. \tag{2.4}$$

He added 'This might well be the *general wave equation* which *no longer* contains the integration constant E, but contains time derivatives instead.' He continued to think about this matter, and five days later, on June 11, in writing to Planck he said 'By the way, during the last few days another heavy stone has been rolled away from my heart... I had considerable anxiety over it... But it all resolved itself with unheard of simplicity and unheard of beauty.' What was this resolution? It was (2.4) above.

Why did Schrödinger not write down simply the correct time dependent equation (1.2) rather than the more complicated (2.4)? He certainly knew the simpler equation but chose to go to the more complicated second order equation*. Why? I suggest the answer is as follows.

Schrödinger did not want his wave equation to contain i, so in a way he eliminated it by going to the fourth order equation (2.4), utilizing $i^2 = -1$. That he tried to avoid i was quite natural since he had started on wave mechanics in Schrödinger (1926*a*) by writing down the real Hamilton–Jacobi equation

$$H\left(q, \frac{\partial S}{\partial q}\right) = E,$$

together with

$$S = K \log \psi.$$

His ψ up to this point was real and time independent. Later in § 3 of Schrödinger (1926*a*) he wrote 'It is, of course, strongly suggested that we should try to connect the function ψ with some *vibration process* in the atom ...'. Alas this was not a simple process, because Schrödinger had to

---

* In all five of Schrödinger's papers written before June 11, 1926 (Schrödinger, 1926*a–e*), (1.2) never appeared. Yet one finds such equations as (2.2), which implies that he knew $i\hbar\dot{\psi}_n = H\psi_n$. The confusing discussion in his June 6 letter to Lorentz about the real part ψ_r of ψ is very revealing in that it shows how Schrödinger was struggling to eliminate/define the imaginary part.

struggle with the question of what frequency to use for this vibration. The subsequent evolution of his thinking on this question is a very interesting topic, but is not the subject we are considering here. What is relevant to us now is the fact that Schrödinger had started his conceptualization of wave mechanics by envisaging a description of a vibration in terms of a real function of space-time. Later when he superposed ψ's he again meant to add real ψ's, each of which depends on time sinusoidally.

To return to Schrödinger's letter of June 11 to Planck, he further emphasized that in (2.4) one may 'let the potential energy be an explicit function of the time.' This turned out to be incorrect and Schrödinger realized this in the next ten days, during which he wrote Schrödinger (1926f) which was received by the publisher on June 23. It was in this paper that the concept was first stated that ψ is a complex function of space-time and satisfies the complex time evolution equation (1.2) which Schrödinger called the *true** wave equation, in contrast to $H\psi = E\psi$ which he called the vibration or amplitude equation.

I should emphasize that I do not infer from the chronology outlined above that Schrödinger's discovery in Schrödinger (1926f), that ψ should be complex, was started by Lorentz's letter of May 27, 1926, to him. That may be the case, but it also could be that after writing Schrödinger (1926d), which was on time independent perturbation theory, Schrödinger began to work on a perturbation theory where the perturbation is time dependent. He would then have to study the time evolution of the wave function ψ, and Lorentz's letter may have arrived in the midst of such a study. What one can be certain is that between June 11 and June 23 Schrödinger was finally convinced that ψ is complex.

A few days after Schrödinger (1926f) was submitted Born submitted the first of his two historic papers on the statistical interpretation of the wave function (Born, 1926a, b). It is interesting to notice that in the first of these two Born used a real wave function,

$$\sin\frac{2\pi}{\lambda}z$$

for the incoming wave and another real wave function

$$\sin k_{nm\tau}(\alpha x + \beta y + \gamma z + \delta)$$

for the scattered wave. Because everything was real, Born did not use 'absolute square' but only 'square' in the famous footnote (added to the first paper in proof) which Pais referred to as follows (Pais, 1986): 'that

* Schrödinger used *eigentlich*, which I translate as *true*. Elsewhere it has been translated as *real*, which is very confusing in the present context.

great novelty, the correct transition probability concept, entered physics by way of a footnote'. It was only in the second paper that Born used complex numbers for the incoming and outgoing waves.

5.3. Complex numbers in Weyl's gauge theory

We have traced above the entry of complex numbers into fundamental physics during the period 1925–6. In fact, several years before that, Schrödinger (1922) had published a most interesting paper entitled 'On a remarkable property of the quantum orbit of one electron', in which he had already mentioned the possibility of introducing an imaginary factor

$$\gamma = -i\hbar \tag{3.1}$$

into Weyl's 1918 gauge theory. He started with Weyl's 'world geometry', i.e. Weyl's 1918 gauge theory of electromagnetism, summarizing Weyl's idea in a *Streckenfaktor*:

$$\exp\left[-\frac{e}{\gamma} \int (V\,dt - \mathbf{A} \cdot \mathbf{dx}) \right], \tag{3.2}$$

and went on to remark that for a hydrogen atom where $A = 0$, the expression in the exponential is equal to

$$-\gamma^{-1} e \bar{V} \tau$$

where τ is the period. For a Bohr orbit with quantum number n this is equal to $-\gamma^{-1} nh$, an integral multiple of $\gamma^{-1}h$. Schrödinger called this result remarkable and said he could not believe that it was without deep physical significance.

At the end of the paper Schrödinger mentioned two possible values for γ, $\gamma = e^2/c$, a real number, or $\gamma = -i\hbar$, i.e. (3.1) above. For the latter case, he remarked, the factor (3.2) becomes unity.

In his great papers of 1926 which created wave mechanics Schrödinger did not refer to this 1922 paper. But Raman and Forman (1969) in their historical research argued that this 1922 paper had in fact played an important role in 'Why was it Schrödinger who developed de Broglie's ideas?' Their thesis was later confirmed by Hanle (1977, 1979; see also Wessels, 1977), who found the following passage in a letter dated November 3, 1925 from Schrödinger to Einstein:

> The de Broglie interpretation of the quantum rules seems to me to be related in some ways to my note in the Zs. f. Phys. 12, 13, 1922, where a remarkable property of the Weyl 'gauge factor' $\exp[-\int \phi\,dx]$ along each quasi-period is shown. The mathematical situation is, as far as I can see, the same, only from me much more formal, less elegant and not really shown generally. Naturally de Broglie's consideration in the framework of his large theory is altogether of

far greater value than my single statement, which I did not know what to make of at first.

Thirteen days later, on November 16, 1925, Schrödinger wrote to Lande (Raman and Forman, 1969):

> Recently I have been deeply involved with Louis de Broglie's ingenious thesis. It's extraordinarily stimulating but nonetheless some of it is very hard to swallow. I have vainly attempted to make myself a picture of the phase wave of an electron in an elliptical orbit. The 'rays' are almost certainly neighboring Kepler ellipses of equal energy. That, however, gives horrible 'caustics' or the like as the wave front. At the same time, the length of the wave ought to be equal to [that of the orbit traced out by the electron in] one Zeeman or Stark cycle!

Schrödinger was by then evidently well on his way to the first great paper on wave mechanics which he submitted on January 27, 1926 (Schrödinger, 1926*a*)!

The Raman–Forman–Hanle thesis that Schrödinger's 1922 paper played an essential role in the creation of wave mechanics is clearly correct. We illustrate this fact by an arrow in Fig. 5.1.

Fig. 5.1. Flow of ideas relating to complex phases and gauge fields. The importance of the 1922 paper of Schrödinger was discovered by Raman and Forman (1969) and Hanle (1977, 1979).

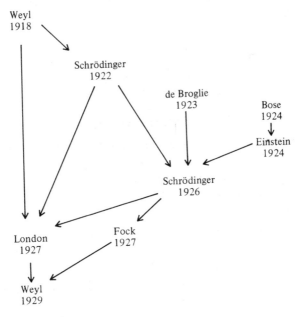

Why then did Schrödinger not refer to his own 1922 paper in 1926? The reason is probably as follows: the 1922 paper discussed a factor (3.2) above, with which (3.1) becomes

$$\exp\left[ei \int A_\mu \, dx^\mu/h \right] \tag{3.3}$$

while the 1926 papers were related to de Broglie's idea which, for comparison, could be put in the form of a factor

$$\exp\left[i \int p \cdot dx/h \right]. \tag{3.4}$$

The two are similar but not the same, and Schrödinger recognized that the relevant one to start wave mechanics from was (3.4) and not (3.3).

While Schrödinger was busy with developing wave mechanics, his 1922 paper caught the attentions of F. London who wrote to Schrödinger a very interesting letter reprinted in Raman and Forman (1969). We attach as an appendix a translation of this letter, which according to Raman and Forman, was written around December 10, 1926.

London developed further this thinking in a paper entitled 'Quantum mechanical meaning of the theory of Weyl' (London, 1927a; see also London, 1927b). A little earlier, Fock had published a paper which discussed invariance of wave equations (Fock, 1927). Both are somewhat confusing* as is natural in those early days of wave mechanics, but both contain the right idea that, in today's notation, electromagnetism enters in wave mechanics with an operator

$$(\partial_\mu - ieA_\mu)$$

on ψ, which is the heart of the gauge principle (see Fig. 1). The definitive discussion of electromagnetism as a gauge theory came later, in an important paper (Weyl, 1929; see also Yang, 1986).

5.4. Modern consequences

The importance of the introduction of complex amplitudes with phases into physicists' description of nature was not fully appreciated until the 1970s when two developments took place: (1) all interactions were found to be some form of gauge field; and (2) gauge fields were found to be related to the mathematical concept of fibre bundles (Wu and Yang, 1975), each fibre being a complex phase or a more general phase. With these developments there

---

* In my article in *Ann. N.Y. Sci.* 294, 86 (1977) I had said that London's paper pointed out the similarity between Fock's work and the 1918 Weyl paper. This is wrong. I had misread the meaning of the footnote on p. 111 of W. Pauli, *Handbuch der Physik* Vol. 24, Part 1 (1933).

arose a basic tenet, of today's physics: *all fundamental forces are phase fields* (Yang, 1983). Thus the almost casual introduction in 1922 by Schrödinger of the imaginary unit i into (3.1) above has flowered into deep concepts that lie at the very foundation of our understanding of the physical world.

In 1975 Wu and I drew up a 'dictionary', reproduced here as Table 1, identifying physicists' terminology for gauge fields with mathematicians' terminology for fibre bundles (Wu and Yang, 1975).

There is in this 'dictionary' a blank space with a question mark because at that time the mathematicians had not studied the concept that corresponds to physicists' 'sources', i.e. density–current four-vector, a natural and fundamental concept in Maxwell's theory of electromagnetism. In the language of the mathematicians, this concept would have been written as

$$* \partial * f = J. \tag{4.1}$$

A sourceless case would satisfy

$$* \partial * f = 0. \tag{4.2}$$

The mathematicians have now studied (4.2) and the results have helped to resolve some deep and long standing problems in topology and differential geometry, providing a modern example, so abundant in past centuries but

Table 5.1. *Reproduction of 'dictionary' comparing terminologies in gauge field theory and fibre bundle theory*

Gauge field terminology	Bundle terminology
gauge (or global gauge)	principal coordinate bundle
gauge type	principal fibre bundle
gauge potential b_μ^k	connection on a principal fibre bundle
S_{ba}	transition function
phase factor Φ_{QP}	parallel displacement
field strength $f_{\mu\nu}^k$	curvature
source^a J_μ^K	?
electromagnetism	connection on a $U_1(1)$ bundle
isotopic spin gauge field	connection on a SU_2 bundle
Dirac's monopole quantization	classification of $U_1(1)$ bundle according to first Chern class
electromagnetism without monopole	connection on a trivial $U_1(1)$ bundle
electromagnetism with monopole	connection on a nontrivial $U_1(1)$ bundle

^{a} i.e. electric source; this is the generalization of the concept of electric charges and currents.

rare now, of how physics could supply powerful insights for the advance of mathematics (Freed and Uhlenbeck, 1984; Lawson, 1985).

5.5. Appendix

*A letter from F. London to E. Schrödinger**
Translated into English by Prof. T. C. Meng.

Dear Professor,

I must have a serious word with you today. Are you acquainted with a certain Mr. Schrödinger, who in the year 1922 (*Zeits. fur Phys.*, 12) described a 'bemerkenswerte Eigenschaft der Quantenbahnen'? Are you acquainted with this man? What! You affirm that you know him very well, that you were even present when he did this work and that you were his accomplice in it? That is absolutely unheard of. So you already know for four years that in the continuous space-time in which atomic processes have to be studied, no rulers and clocks can be used to define an Einstien–Riemann metrical relationship (*Masszausammenhang*), once one has to see whether the general metrical principles which have been expressed by Weyl's theory of distance transference (*Streckenübertragung*) is perhaps helpful. And you have for four years very well noticed that they are even extremely helpful. Namely while usually nonsense emerges by using Weyl's distance transference [Einstein's objection (Yang, 1986), Weyl's very poor excuse (Weyl, 1968) with 'adjustment' (*Einstellung*)] you have shown that on the discrete physical orbits the scale unit (*Eicheinheit*) (with $\gamma = 2\pi i/h$) can be reproduced for spatially closed paths; and in fact you observed at that time for the nth orbit the scale unit swells and shrinks (*anschwillt und zusammenschrumpft*) exactly n times, just like the standing wave which describes the location of the charge. So you have shown that the theory of Weyl is only reasonable – i.e., it leads to a unique measure-determination – when one combines this theory with the quantum theory. Actually there is nothing else one can do, if the entire atomic world is a continuous space-time without any fixed point for identification. You knew this and said nothing and made no statement about it. This kind of thing has never happened before. You wrote very modestly in your paper (p. 14): You did not – to confess immediately – come very far in the discussion of the possible meaning of this fact. But in this paper you not only have put the hopeless confusion of the Weyl theory to an end, but also even had the resonance character of the quantum postulate in your hands long before de Broglie, and also thought about whether you should take $\gamma = h/2\pi i$ or $e^2/c!$ (p. 23) – Will you now immediately confess that, like a priest, you kept secret the truth which you held in your hands, and give notice to your contemporaries of all you know! The

* According to Raman and Forman, who reprinted this letter in their article in *Historical Studies in the Physical Sciences*, Vol. 1, pp. 291–314, the letter was written around December 10, 1926.

most important thing is yet to be done, that remark made in 1922 being a theorem of the old quantum mechanics. One can with certainty expect that it will show its whole significance when it is brought into meaningful connection with wave mechanics. (I have not done this yet.) I think that it is your duty, after you have mystified the world in such a manner, now to clarify everything.

Now, that is enough. Thank you very much for spending so much time on my stupid letter*. For the moment I have discontinued my study on this matter. I think on the whole the Kaluza–Klein Space Theory has to be considered as a set-back (*Rückschritt*) since the existence of the beautiful Weyl Space Theory, and I would like to look at this more closely. I have different clues (*Anhalt*) which show that it will not be difficult to make Weyl's and Kaluza's theory consistent with each other (plot for yourself for every world-point the scale unit (*Eicheinheit*) as the 5th dimension, one immediately sees a lot of beautiful things!) I am eagerly looking forward to reading your manuscript† (until now it is still not here) especially after the hints given by Fues. Even if it would be just for one day, I would very, very much like to be able to see it.

By the way, the Rockefeller is granted; the telegram arrived yesterday. I am very happy that it is now certain that I may work with you.

I wish you a good journey. I am looking forward to your return.

With hearty greetings, I am

Yours very faithfully

Fritz London

References

Born, M. (1926*a*) *Z. f. Phys.* 37, 863 (received June 25)

Born, M. (1926*b*) *Z. f. Phys.* 38, 803 (received July 21)

Born, M. and Jordan, P. (1925) *Z. f. Phys.* 34, 858 (received September 27)

Dirac, P. A. M. (1925) *Proc. Roy. Soc.* A109, 642 (received November 7)

Dirac, P. A. M. (1972) *Fields & Quanta* 3, 139

Freed, D. and Uhlenbeck, K. (1984) *Instantons and Four-manifolds*. Springer

Fock, V. (1927) *Z. f. Phys.* 39, 226

Hanle, P. (1977) *Isis* 68, 606

Hanle, P. (1979) *Am. J. Phys.* 47, 644

Heisenberg, W. (1925) *Z. f. Phys.* 33, 879 (received July 29)

Lawson, Jr, H. B. (1985) *The Theory of Gauge Fields in Four Dimensions*. American Mathematics Society 58, Providence, RI

London, F. (1927*a*) *Z. f. Phys.* 42, 375

London, F. (1927*b*) *Naturwiss.* 15, 187

Pais, A. (1986) *Inward Bound*, p. 257. Oxford University Press

* Raman and Forman comment that this refers to a letter of December 1, 1926, to which Schrödinger replied on December 7, 1926.

† Raman and Forman comment that this refers to a manuscript on relativistic wave equations mentioned in Schrödinger's letter of December 7, 1926.

Raman, V. V. and Forman, P. (1969) *Hist. Studies Phys. Sci.* 1, 291
Schrödinger, E. (1922) *Z. f. Phys.* 12, 13
Schrödinger, E. (1926a) *Ann. d. Phys.* 79, 361 (received January 27)
Schrödinger, E. (1926b) *Ann. d. Phys.* 79, 489 (received February 23)
Schrödinger, E. (1926c) *Ann. d. Phys.* 79, 734 (received March 18)
Schrödinger, E. (1926d) *Ann. d. Phys.* 80, 437 (received May 10)
Schrödinger, E. (1926e) *Die Naturw.* 28, 664
Schrödinger, E. (1926f) *Ann. d. Phys.* 81, 109 (received June 23)
Wessels, L. (1977) *Studies Hist. & Phil. Sci.* 10, 311
Weyl, H. (1929) *Z. f. Phys.* 56, 330
Weyl, H. (1968) in (Chandrasekharan, K., ed.) *Gesammelte Abhandlungen*, vol. II, p. 261. Springer
Wu, T. T. and Yang, C. N. (1975) *Phys. Rev. D* 12, 3845
Yang, C. N. (1983) *Selected Papers 1945–1980 with Commentary*, p. 564. Freeman
Yang, C. N. (1987) (Chandrasekharan, K., ed.). Springer (in press)

6

Consequences of the Schrödinger equation for atomic and molecular physics

W. E. THIRRING

Technische Universität Wien

6.1. Introduction

It was a major breakthrough when Schrödinger proposed his equation

$$i \frac{\partial \psi}{\partial t} = H \psi. \tag{1.1}$$

It turned out to be of much wider validity than the special case considered by him, namely electrons and nuclei. But this alone explained according to Dirac's famous phrase 'all of chemistry and most of physics'. In this situation the Hamiltonian is

$$H = \sum_{i=1}^{N} \frac{p_i^2}{2m_i} + \sum_{i>j} \frac{e_i e_j}{|x_i - x_j|}, \tag{1.2}$$

where (x_i, p_i, m_i, e_i) are coordinate, momentum, mass and electric charge of particle i, $i = 1, \ldots, N$, and p_i^2 is to be interpreted as the differential operator $-\hbar^2 \Delta_i$. This equation should describe atoms and molecules of arbitrary complexity up to macroscopic bodies. Small wonder that legions of physicists tried to extract from these equations information about the strangest properties of odd substances using all thinkable legitimate and illegitimate mathematical means. In these heroic attempts they were not discouraged by the failure of mathematicians of making much headway with the supposedly simpler classical problem posed by (1.2). In fact, the greatest minds got stuck already at $N = 3$ and gave up in despair. In spite of more than one century's efforts the classical three-body problem still offers many unanswered questions. To one's surprise quantum theoreticians actually got much further with the Schrödinger equation. Of course it would be pretentious to claim that one can deduce the most subtle effects of complex systems from (1.1) and (1.2). But it is all the more gratifying that modern functional analysis supplies the tools to deduce at least the main qualitative features of atoms and molecules from the Schrödinger equation.

To cut this review of the work of many people during half a century to a reasonable size I have to impose myself several limitations:

(1) I shall only consider (1.2) and ignore other interactions, relativistic effects, etc. Also I shall not substitute mesons or antiparticles in the place of nuclei to have more variations of the parameters.

(2) I shall concentrate on results pertaining to the main qualitative features of atoms and molecules and disregard finer details.

(3) By results I mean general rules which can be deduced from (1.2) without further assumptions or uncontrollable approximations. To make this clear I shall call them theorems but I will not attempt to give the proofs. For these I have to refer the reader to the references. I will only comment on the meaning and domains of applicability of these theorems.

(4) A complete list of references would cover volumes. I shall only give a (necessarily biased) selection of references for the cases at hand. For this I have to apologize to all the unquoted authors.

Some people think that physics is over once the equation is found which governs some phenomena. To me this seems as foolish as somebody who says English is over once he has learned the words and the grammar and never goes on to read and understand Shakespeare. Physics is not the equation but the multitude of phenomena which result from it. To know the equation is not the end but the beginning and to deduce from it the physics is an unending quest. The following is a bird's-eye view of the road covered up to now.

6.2. Self-adjointness of H

After having settled the questions of existence and uniqueness of the solution of the equations of motion of a dynamical system, one has to ask whether the time evolution can be extended to infinite times or whether some trouble arises at a finite time. What can go wrong for classical particles is that they might fall into a singularity or reach infinity after a finite time. Actually for classical Coulomb systems both possibilities do happen. For $N = 3$ the orbits which lead to collisions may fill large portions in phase-space and for $N = 4$ (Moser, 1973; Thirring, 1978) there are orbits where one particle is kicked to infinity in a finite time. In quantum mechanics most physicists ignore these difficulties and formally solve (7.1) by $\psi(t) = e^{-iHt}\psi(0)$. They believe that a hermitian Hamiltonian H generates a unitary time evolution e^{-iHt} for all real t. Though this belief is false it is true that these difficulties of classical Coulomb systems disappear miraculously in quantum mechanics. Since this remarkable situation is generally not appreciated it may be worth a few lines to state the fact (Reed, 1975; Thirring, 1981*a*). For a bounded hermitian Hamiltonian e^{-iHt} is indeed unitary for all t. However, H from (1.2) is unbounded, only some vectors from the Hilbert space $\mathcal{H}$ are not made infinitely long by H. They form $D(H) \subset \mathcal{H}$, the domain of definition of H (which will be dense in $\mathcal{H}$ in the cases to be considered). H is called hermitian if $\langle \psi | H\phi \rangle = \langle H\psi | \phi \rangle$ for all ψ, $\phi \in D(H)$ and it is called self-adjoint if in addition $D(H) = D(H^*)$, where $D(H^*)$ are the vectors ψ such that $|\langle \psi | H\phi \rangle| \leqslant M < \infty$ for all $\phi \in D(H)$, with $\langle \phi | \phi \rangle = 1$. H^* is defined on $D(H^*)$ by $\langle H^*\psi | \phi \rangle = \langle \psi | H\phi \rangle$ and the above condition assures $\| H^*\psi \| \leqslant M$. The key result about the question when the time evolution goes on forever is

Theorem (2.1)
e^{-iHt} is unitary if and only if H is self-adjoint.

Remarks (2.2)
(1) To see what goes wrong if H is only hermitian but not self-adjoint consider the shift on $\mathcal{H} = L^2((0, \infty))$. It is generated by

$$p_r = \frac{1}{i} \frac{\partial}{\partial r}$$

which is hermitian on $D(p_r) = \{\psi \in \mathcal{H}, \psi(r=0) = 0 \text{ and } p_r\psi \in \mathcal{H}\}$. However, $U(\alpha) = \exp i\alpha p_r$ cannot be unitary since

$$\| U(\alpha)\psi \| = \int_0^\infty dr |\psi(r+\alpha)|^2 = \int_\alpha^\infty dr |\psi(r)|^2 < \int_0^\infty dr |\psi(r)|^2.$$

By shifting the wave function towards the origin one loses normalization and thus the unitarity.
(2) It may happen that $D(H)$ was chosen too small and that H can be 'extended' to a self-adjoint operator. This means that a normal Hamiltonian does not yet specify the time evolution but only together with appropriate boundary conditions. A hermitian H may have many self-adjoint extensions or none (like in the previous example).
(3) Even if T and V are self-adjoint, $T + V$ is not. It is *a priori* defined on $D(T) \cap D(V)$ and there it is hermitian but in general not self-adjoint.

Having been confronted with these facts of life one might wonder whether Schrödinger by writing down a formal Hamiltonian without specifying its domain has actually predicted everything about atoms and molecules. That he has actually done so comes from the fact that the Coulomb potential, though unbounded, is small compared with the kinetic energy.

Theorem (2.3)

If T is self-adjoint and V hermitian on $D(V) \supset D(T)$ and $\|V\psi\| \leqslant a\|\psi\| + b\|T\psi\|$ for $a < \infty$, $b < 1$, and $\psi \in D(T)$, then $T + V$ is self-adjoint on $D(T)$.

Theorem (2.4)

If

$$T = \sum \frac{p_i^2}{2m_i}$$

and

$$V = \sum_{i>j} \frac{e_i e_j}{|x_i - x_j|}$$

then for all $b > 0$ there is an $a < \infty$ such that $\|V\psi\| \leqslant a\|\psi\| + b\|T\psi\|$.

These two theorems clear away all the clouds of ghastly classical pathologies and $T + V$ is self-adjoint for all values of $m_i > 0$, e_i and N.

Remarks (2.5)

(1) (2.4) is the mathematical version of the following folklore with the uncertainty principle. To make the Coulomb singularity $-1/r$ dangerous the wave function has to be concentrated in a small region Δr but then the kinetic energy is larger than $(\Delta p)^2 > 1/(\Delta r)^2$ and thus becomes more positive than the potential $-1/\Delta r$ is negative. Thus in quantum theory the Coulomb singularity is harmless.

(2) For a potential γ/r^2 there are for $\gamma < 0$ regions in phase space of infinite volume (for $l^2 + \gamma \leqslant 0$) for which the classical orbit of one particle spirals into the singularity. In quantum theory $p^2 + \gamma/r^2$ is no longer self-adjoint on $D(p^2) \cap D(1/r^2)$ already for $\gamma < 3/4$ (Narnhofer, 1974).

(3) For the relativistic expression $(p^2 + m^2)^{1/2}$ instead of $p^2/2m$ (2.4) is no longer true, b cannot be made arbitrarily small and $T + V$ for $N = 2$ is only self-adjoint if $e^2 < 1/2$. For $e^2 > 1/2$ there are self-adjoint and not self-adjoint extensions. They describe the particle falling into the singularity and bouncing back or disappearing at $r = 0$.

(4) One might argue that in reality the $1/r$ singularity is smeared out by nuclear seize effects and all this mathematical sophistication is irrelevant since for finite potentials there is no problem. Though this objection is correct, what (2.3) and (2.4) show is that even in the limit of zero nuclear radius there is a well-defined time evolution which is insensitive to the nuclear structure. Thus we can make atomic physics without knowing nuclear physics.

6.3. The spectrum of H

Roughly speaking the spectral values are those (real or complex) numbers which an operator can assume. Formally they are given by

Definition (3.1)
The spectral values z of a (bounded or unbounded) operator H are those for which the operator $H - z$ is not invertible.

Remarks (3.2)
(1) In finite dimensions hermiticity and self-adjointness are synonymous and there the spectral values of a hermitian operator are its eigenvalues which are real.
(2) Since H (and $H - z$) map $D(H)$ into $\mathscr{H}$ if $(H - z)^{-1}$ exists it maps $\mathscr{H}$ onto $D(H)$. It may fail to do so either because $H - z$ is not $1 - 1$ (in which case z is an eigenvalue) or because $(H - z)D(H)$ is not all of $\mathscr{H}$. This happens for $\operatorname{Im} z \neq 0$ if H is hermitian but not self-adjoint. If H is self-adjoint its spectrum is part of the real axis.

A hermitian matrix can be diagonalized which means that by a unitary transformation we can make it a multiplication operator which multiplies the eigenvectors by the eigenvalues. For self-adjoint operators this fact generalizes to the famous Reed (1972) and Thirring (1981a).

Spectral theorem (3.3)
Given a self-adjoint operator H, $\mathscr{H}$ is unitarily equivalent to a sum of $\mathscr{H}_i$ where H acts in each $\mathscr{H}_i$ as a multiplication operator, $(H\psi)(h) = h\psi(h)$, and in $\mathscr{H}_i$ the scalar product is given by $\langle \psi | \phi \rangle = \int d\mu_i(h)\psi^*(h)\phi(h)$, where $d\mu(h)$ is some measure over the spectrum of H.

Remarks (3.4)
(1) The measure may be concentrated on some points h_j (i.e. of the form $\sum_j \delta(h - h_j)dh$). Then one says $\mathscr{H}_i$ belongs to the *discrete spectrum* of H and the h_j are the eigenvalues. If H is the Hamiltonian then a vector concentrated on one of the h_j's is a bound state in which particles stay infinitely long in a bounded region.
(2) If $d\mu_i$ is of the form $f_i(h)dh$, where $f_i(h)$ is some (positive) integrable function, one says $\mathscr{H}_i$ belongs to the *absolutely continuous spectrum*. Vectors from this $\mathscr{H}_i$ for the Hamiltonian are states where the particle leaves any bounded region in some finite time.
(3) There is a third possibility where $d\mu_i(h)$ is concentrated on a set of

Lebesgue measure zero but not on discrete points. This $\mathscr{H}_i$ is called the *singular continuous spectrum*. It corresponds to states where particles wander arbitrarily far away but may keep coming back. This situation was for a long time considered to be a pathology but now seems to be an important part of the quantum theory of disordered systems (Daubechies, 1983).

Though classically already in the restricted three-body problem there are orbits where a particle moves away for arbitrarily long times but keeps coming back (Moser, 1973) in quantum theory we have

Theorem (3.5)
For the Coulomb Hamiltonian (1.2) the continuous singular spectrum is absent.

To discuss the spectrum of H we have therefore only to consider the discrete and the absolutely continuous part.

Theorem (2.4) has told us that in some sense the Coulomb potential is always weaker than the kinetic energy and thus non-relativistically atoms should not collapse. Actually one can easily (Thirring, 1981a) deduce from (2.4)

Theorem (3.6)
H is bounded from below (for all Z and N). This means that the spectrum of H is located between some $E_0 \leqslant 0$ and ∞.

Remarks (3.7)
(1) If one takes the relativistic kinetic energy $(p^2+m^2)^{1/2}$ (3.6) is false. For Ze^2 sufficiently large the spectrum of H goes down to $-\infty$. This means that then the electrons fall into the nuclei.
(2) Theorem (3.6) is true irrespective of the statistics of the electrons, though the value of E_0 would be much lower if the electrons were bosons.

6.4. The discrete spectrum

We shall consider here only the atomic case (one nucleus with mass $= \infty$, N electrons) and return to the molecular case in § 6.6. By a dilation one can transform (Thirring, 1981a) this Hamiltonian into

$$H(N,\alpha)=\frac{1}{2}\sum_{i=1}^{N} p_i^2 - \sum_{i=1}^{N}\frac{1}{|x_i|}+\alpha\sum_{i>j}\frac{1}{|x_i-x_j|}. \qquad (4.1)$$

We have chosen atomic units so that H contains only one parameter

$\alpha = 1/Z$. For $N\alpha < 1$, 1 or > 1 we have a positive ion, a neutral atom or a negative ion.

What limits the point spectrum from the other side of (3.6) is the

Virial theorem (4.2)

If ψ is a (normalized) eigenfunction of H, $H\psi = e\psi$, then

$$e = -\langle \psi | \tfrac{1}{2} \sum p_i^2 | \psi \rangle.$$

Corollary (4.3)

Since $p_i^2 > 0$ the point spectrum is located between $E_0(N, \alpha)$ and 0.

Remark (4.4)

From the proof of (4.2) one actually learns a little bit more, namely that e plus the expectation value of the kinetic energy equals the expectation value of $d^2/dt^2 \tfrac{1}{2} \sum x_i^2$. If $e > 0$ then $\sum x_i^2$ must go for large t like t^2 and one particle has to keep going to infinity in accordance with (3.5).

The next question is where does the continuous spectrum start. One would expect it to start at an energy where one electron can go to infinity, that is the ionization threshold. This expectation is confirmed by Hunziker (1966), van Winter (1964) and Zhislin (1960).

Theorem (4.5)

The continuous spectrum of $H(N, \alpha)$ starts at $E_0(N - 1, \alpha)$.

Actually if one electron runs off to infinity it leaves behind an atom with one electron less and this has a net positive charge $Z_n = 1/\alpha - (N - 1)$. If $Z_n > 0$ then the electron should feel as if it is in a hydrogen atom with charge Z_n and be able to form infinitely many bound states. Nevertheless it took decades to prove (Kato, 1951)

Theorem (4.6)

For $\alpha < 1/N - 1$ there are infinitely many eigenvalues between $E_0(N, \alpha)$ and $E_0(N - 1, \alpha)$.

One might think that these are all the eigenvalues since in the hydrogen atom there are none imbedded in the continuum. There this is excluded by (4.3) but already for $N = 2$ (4.3) would allow eigenvalues between $E_0(1, \alpha)$ and 0. Actually there are plenty.

Theorem (4.7)

For $\alpha < 1$ there are infinitely many eigenvalues of $H(2, \alpha)$ imbedded in the continuum between $E_0(1, \alpha)$ and 0.

Remark (4.8)

Theorem (4.7) does not seem to be common knowledge since people are fascinated by the trivial decomposition of the helium spectrum into ortho- and para-helium but are not aware of the more subtle distinction between states of natural and unnatural parity. Thus I will pause to explain (4.6). For $\alpha = 0$ it is obvious since the eigenvalues of $H(2,0)$ are the sum of eigenvalues of $H(1,0)$ and $H(1,0)$. $H(1,0)$ is hydrogen and Balmer has taught us

$$e_n = -\frac{1}{2}\frac{1}{n^2}.$$

However, $e_1 + e_2 = -1/4$ which is above the ionization threshold $e_1 + e_\infty = -1/2$. If α is turned on the Coulomb repulsion will actually make an Auger transition in the $e_2 + e_2$ state, one electron falling to e_1, the other one being kicked out. However, it may happen that this is forbidden by conservation laws. If the $n=2$ states are p-states which combine to a $L=1$ state we still have positive parity. However, if one electron is in the ground state ($l=0$, parity $+1$) the parity of such a state is $(-)^L$ and cannot be reached from $L=1$, parity $+1$. Thus the 1^+-state remains stable and there are infinitely many of them. Of course these states are stable only within the theory determined by the Hamiltonian (4.1). They will decay by radiative transition or through external influences but the width owing to these possibilities is much smaller than the one of the other Auger states.

If $\alpha = 1$ and $N = 2(=H^-)$ there is no easy general theorem about the discrete spectrum and it took great ingenuity to prove (Grosse and Pittner, 1983; Hill, 1977)

Theorem (4.9)
$H(2, 1)$ has exactly two eigenvalues, one isolated $< -1/2$ and one in the continuum $(-1/2, 0]$. (The latter being 3-fold degenerate.)

One might wonder how many excess electrons a large atom can bind. There are two results

Theorem (4.10) (Lieb, 1984)
$H(N, \alpha)$ has no discrete spectrum for $N \geq (1 + 2/\alpha)$.

Theorem (4.11) (Lieb, Sigal and Simon, 1984)
If the electrons obey Fermi statistics the $\alpha(N)$ for which the discrete spectrum disappears obeys $\lim_{N \to \infty} N\alpha(N) = 1$.

Remarks (4.12)
(1) Whereas (4.10) also holds for bosons, (4.11) reads for bosons (Baumgartner, 1984; Benguria and Lieb, 1983) $\lim_{N \to \infty} N\alpha(N) = 1.21$. Thus, if the electrons were bosons, a uranium atom could bind about 20 excess electrons.
(2) Theorem (4.10) has in this form the best possible constant, since for $\alpha = 1$, $N = 2$ ($= H^-$) there are four bound states. However, one might think that there are no doubly negatively charged ions and (4.10) should read $N \geqslant (2 + 1/\alpha)$. We have seen that for bosons this is false but for fermions this conjecture is still tenable but hard to prove.

The exact location of the eigenvalues and in particular the ground state energy $E_0(N, \alpha)$ is a formidable numerical problem. Generally one knows that $E_0(N, \alpha)$ cannot be a wildly fluctuating function.

Theorem (4.13) (Narnofer and Thirring, 1975; Rebane, 1973)
$-(|E_0(N, \alpha)|)^{1/2}$ is in α concave.

Remark (4.14)
Since $E_0(N, \alpha) = \min_{\langle \psi | \psi \rangle = 1} \langle \psi | H(N, \alpha) | \psi \rangle$ and H is linear in α it is trivial that $E_0(N, \alpha)$ is in α concave since any lower envelope of linear functions is concave. The stronger statement (4.12) goes about as far as possible. If one plots $E_0(N, \alpha)$ it is still curved downward, where $-(|E_0(N, \alpha)|)^{1/2}$ is almost a straight line.

Since $E_0(N, \alpha)$ cannot be determined analytically but only approximately one needs good error estimates to make an honest statement. With the variational principle and modern computer technology one can find upper bounds with sufficient accuracy. Sufficiently means that the theoretical error is less than other corrections to H not contained in (4.1) Unfortunately the situation for lower bounds is not so good. For $N = 2, 3$ there are projection techniques which have also sufficient accuracy so there the problem can be considered as solved (Thirring, 1981a; Weinstein and Stenger, 1972). For instance the theoretical error bars are small enough so that Hund's rule is born out. For larger N these methods become too clumsy and one only has an effective field theory which gives lower bounds with an error of a few per cent (Hertel, Lieb and Thirring, 1975) like the Hartree–Fock energies as upper bounds. But since these are percentages of the total energy which is huge ($\sim$MeV for uranium) these methods are not fine enough to reveal features like the shell structure of atoms and the periodic table.

6.5. The continuous spectrum

The explanation of some scattering experiments was the first genuine triumph for Schrödinger's theory because at that time it was thought that Heisenberg's matrix mechanics was not applicable to such problems and the Schrödinger equation had to be used. In the course of time one has found out that scattering theory is a far more wider scheme and not only a speciality of Schrödinger's equation. Scattering theory classifies dynamical systems by assigning them to equivalence classes of systems with equal asymptotic behaviour. Systems in one class are isomorphic in two ways depending on whether one identifies them at $t = +\infty$ or $t = -\infty$. The scattering transformation is the composition of one of these isomorphisms with the inverse of the other. The first question about the absolutely continuous spectrum is whether it belongs to the equivalence class of free particles. This means that for $t \to +\infty$ some particles escape to infinity and approach a free orbit. This is not necessarily so and counterexamples can be constructed where a particle comes from infinity but then gets stuck and does not return to infinity. Classically such orbits exist, but if $H = $ kinetic energy + reasonable potential then they cover in phase space only a set of measure zero. There are quantum mechanical systems (Pearson, 1975), where a wildly oscillating potential keeps a wave packet from diffusing out again.

In scattering theory one compares the motion governed by a Hamiltonian H with the one of a free particle (Hamiltonian H_0). In quantum mechanics one then asks whether the limits

$$\lim_{t \to \pm\infty} e^{iHt} e^{-iH_0 t} \equiv \Omega_\pm \tag{5.1}$$

exist. The $\Omega_\pm$ map unitarily the free motion into the motion with interactions and the scattering transformation is

$$S = \Omega_+^{-1} \Omega_-. \tag{5.2}$$

Remarks (5.3)

(1) One of the main points of scattering theory is the seemingly technical question about the operator topology in which the limits $\Omega_\pm$ exist (Reed and Simon, 1979; Thirring, 1981a).

(2) The choice of H_0 is somewhat arbitrary and dictated by the experimental setup. For two particles H_0 will be the kinetic energy only. For several particles one may be looking at a channel where some particles are still bound together and H_0 has then to include some interactions.

(3) $\Omega_\pm$ are not unitary maps of the Hilbert space $\mathcal{H}$ into itself but into the part of $\mathcal{H}$ corresponding to the continuous spectrum of H. If it is onto this

part then this means that all vectors in the continuous spectrum approach for $t \to -\infty$ and $t \to \infty$ the free time evolution. In this case one calls the system asymptotically complete, all states either remain in bounded regions or come and go to infinity.

When applying this program to the Coulomb potential one meets already for two particles the difficulty that $\exp(iHt) \exp(-iH_0t)$ converges weakly to zero (Dollard, 1961). The problem appears already at the classical level since for the Kepler problem x does not go asymptotically like $x - pt$ but there is an extra term $\sim \ln t$. To get a useful result one can modify H_0 so as to reproduce this asymptotic behaviour at the expense of having H_0 explicitly depending on t:

$$H_0(t) = \frac{p^2}{2m} + \frac{m\alpha}{|\mathbf{p}|} \frac{\ln |t|}{t}.$$ (5.4)

Then e^{-iH_0t} has to be replaced by $U_0(t)$, the solution of

$$i \frac{dU_0(t)}{dt} = H_0(t) U_0(t)$$

and one finds that the limits

$$\Omega_\pm = \lim_{t \to \pm\infty} e^{iHt} U_0(t)$$ (5.5)

exist in the strong operator topology. Diagonalizing $S = \Omega_+^{-1} \Omega_-$ one recovers the usual Coulomb phase shifts $\delta_l(k)$.

Remark (5.6)
If one is just interested in the scattering angle, i.e. the difference between $\mathbf{p}_+ = \lim_{t \to \infty} \mathbf{p}(t)$ and $\mathbf{p}_- = \lim_{t \to -\infty} \mathbf{p}(t)$ one does not have to choose a particular $H_0(t)$ but can calculate (Grosse *et al.*, 1974) the scattering angle using only the constancy of the Lenz–Runge–Laplace vector

$$\mathbf{F} = \frac{1}{2} (\mathbf{p} \wedge \mathbf{L} - \mathbf{L} \wedge \mathbf{p}) + m\alpha \frac{\mathbf{x}}{|\mathbf{x}|}.$$

However, S does not only transform $\mathbf{p}_-$ into $\mathbf{p}_+ = S^{-1} \mathbf{p}_- S$ but contains also information on the time delay $\partial \delta_l(k)/\partial k$. This is the difference in the time which orbits with H and H_0 spend in a large sphere. It would be infinite for $H_0 = p^2/2m$ but is finite with (5.4) (Narnhofer and Thirring, 1980).

In this sense the quantum mechanical two-body Coulomb problem is asymptotically complete. The same problem for several particles is far more complicated, and even for short range potentials only recently the solution has been announced (Sigal and Soffer, 1986). That this problem has to be hard is illustrated by the following theorem for the classical restricted three-body problem with $1/r$-potentials (Moser, 1973). There exists a time τ so that for any sequence $\tau_i > \tau$ one can find an orbit which returns at the times $\sum_{i=1}^{n} \tau_i$. Thus there are many orbits which neither stay in bounded regions nor escape for good to infinity.

Having settled the general structure of the orbits one wants to have some numerical information about S. Whereas for two particles S can be constructed explicitly it is notoriously hard to get reliable numbers already for $N = 3$. To understand the reason consider scattering of $e^{\pm}$ at low energies on hydrogen. The scattering length (which is essentially the square root of the cross-section) depends sensitively on the existence of low energy bound states. In fact, it becomes infinite if there is a bound state at the beginning of the continuous spectrum. For e^- there is such a loosely bound state, namely the H^- ion, whereas for e^+ the state is presumably just not bound. It is clear that under these circumstances it is very difficult to pin down the scattering length. Once the possibility of a bound state is removed by making the atom sufficiently small (for instance by taking $p\mu^-$ instead of pe^-) one can give fairly tight bounds for the scattering length (Grosse, Narnhofer and Thirring, 1979).

6.6. The limit of infinite nuclear mass

In (1.2) there is one small dimensionless parameter, namely the ratio $m/M =$ electron mass/nuclear mass. It ranges between about 0.5×10^{-3} to 0.25×10^{-5}. In § 6.4 we have already put m/M equal to zero but generally it needs some discussion to neglect the kinetic energy, after all it saved the self-adjointness. In any case it is positive so by throwing it away we underestimate the energy of an atom, $E_0(0) \leqslant E_0(m/M)$. Furthermore, according to (4.14) it is a concave function of m/M and must be rather well-behaved. In fact, the statement can be strengthened to (Narnhofer and Thirring, 1975)

Theorem (6.1)

If E_0 is the ground state energy of an atom then $-1/E_0(m/M)$ is concave in m/M.

Since a concave function is below its tangents one can easily draw from (6.1) the

Corollary (6.2)

$E_0(0) \leqslant E_0(m/M) \leqslant E_0(0)(1 + Nm/M)$.

Remarks (6.3)

(1) For $N = 1$ one knows that

$$E_0\left(\frac{m}{M}\right) = -\frac{1}{2}\frac{Z^2 e^4}{\hbar^2}\frac{m}{1+m/M}.$$

Corollary (6.2) tells us that the reduced mass correction with $m \to Nm$ is the worst thing which can happen for N electrons.

(2) Although (6.2) shows that the influence of the nuclear motion is small for normal atoms, it is not fine enough for exotic atoms. It allows to conclude from the binding energy of H^- that $e^-\mu^+e^-$ is bound but not that $e^-e^+e^-$ is bound.

Born and Oppenheimer proposed that for molecules it was useful to consider the limit of infinite nuclear mass, and indeed a good part of theoretical chemistry rests on this assumption. More precisely, the prescription of Born and Oppenheimer is the following. First separate (1.2) into nuclear and electronic parts (N = number of electrons, $\mathcal{N}$ = number of nuclei, (x_i, p_i) = electron variables, (X_i, P_i) = nuclear variables)

$$H = \sum_{i=1}^{N}\frac{p_i^2}{2m} + \sum_{k=1}^{\mathcal{N}}\frac{P_k^2}{2M_k} - \sum_{i=1}^{N}\sum_{k=1}^{\mathcal{N}}\frac{e^2 Z_k}{|x_i - X_k|} + e^2\sum_{i>j}^{\mathcal{N}}\frac{1}{|x_i - x_j|} + e^2\sum_{k>l}^{\mathcal{N}}\frac{Z_k Z_l}{|X_k - X_l|}$$

$$= H_{el} + \sum_{k=1}^{\mathcal{N}}\frac{P_k^2}{2M_k} + e^2\sum_{k>l}^{\mathcal{N}}\frac{Z_k Z_l}{|X_k - X_l|}, \tag{6.4}$$

next determine the ground state energy $E(X)$ of H_{el} as a function of the nuclear coordinates $X = (X_1, \ldots, X_{\mathcal{N}})$ and then solve for the nuclear motion by looking for the ground state of the Schrödinger equation

$$\left(\sum_{k=1}^{\mathcal{N}}\frac{P_k^2}{2M_k} + E(X) + e^2\sum_{k>l}^{\mathcal{N}}\frac{Z_k Z_l}{|X_k - X_l|}\right)\psi = E_{BO}\psi. \tag{6.5}$$

Born and Oppenheimer argued intuitively by saying that the electrons move so much faster than the nuclei that the latter should appear to them as static. This intuition was actually very good, for the ground state energy E of H (6.4) one has

Thoerem (6.6)

$$E_{BO} \leqslant E_0 \leqslant E_{BO} + \sum_{k=1}^{\mathcal{N}}\frac{c_k}{2M_k}$$

where the $c_k > 0$ are independent of the M_k.

Remark (6.7)
Whereas for the energy the error is only $O(m/M)$, for the wave function

(Combes, 1977) it is $O((m/M)^{1/4})$. Since $1/M$ has the coefficient $\hbar^2$ this shows that the wave function cannot be expanded in $\hbar$ but only in $\hbar^{1/2}$.

Since the Born–Oppenheimer scheme is on solid ground one can formulate the questions why and how atoms attract each other to form molecules. First one notes that the electrons like the nuclei to be close together and it is only the Coulomb repulsion of the nuclei which keeps them apart (Thirring, 1981a).

Thoerem (6.8)
$E(0) \leqslant E(X)$. For $\mathcal{N} = 2$ and $\rho =$ the maximum of the electron density if $X_1 = X_2$ one has the other bound

$$E(X) \leqslant E(0) + \frac{(Z_1 + Z_2)\pi}{6} |X_1 - X_2|^2 \rho.$$

Remarks (6.9)
(1) The upper bound shows that at small distances the nuclear repulsion is always stronger than the attraction due to the electrons. This is no longer the case if one uses the relativistic kinetic energy, then the electron attraction may go as $(X_1 - X_2)^{-1}$ (Daubechies and Lieb, 1983).
(2) For $N = 1$ $E(\lambda X)$ is in λ monotonically increasing (Hoffman-Ostenhof, 1980; Lieb, 1982; Lieb and Simon, 1978). For $N > 1$ one only knows that $\lambda^2 E(\lambda X)$ is in λ concave (Narnhofer and Thirring, 1975).
(3) At the equilibrium position (if it exists) one has

$$\frac{\partial}{\partial \lambda} \left[E(\lambda X) + \frac{e^2}{\lambda} \sum_{k > l} \frac{Z_k Z_l}{|X_k - X_k|} \right] = 0$$

and the virial theorem holds in the form that the kinetic energy of the electrons $=$ |total energy|. In chemical folklore one sometimes comes across the statement that the chemical binding arises because the electrons have more space to move around in the molecule than in the separated atom. Thus one argues that one can save kinetic energy by uniting the atoms. This explanation is false inasmuch as the virial theorem tells us that the kinetic energy of the electrons in the molecule has to be larger than in the separated atom if one gains energy by forming the molecule.

Since the nuclear repulsion wins at small distances the question arises whether at some larger distance $E(X) + e^2 \sum Z_k Z_l / |X_k - X_l| < 0$. Generally this need not be the case, positive ions will repel each other at any distance. However, neutral objects will always attract each other by the van der Waals force, quite independent of their internal structure (Lieb and

Thirring, 1986). By 'object' we mean an atom or a molecule of arbitrary complexity.

Theorem (6.10)

Denote the nuclear coordinates of molecule 1 by $X_1, \ldots, X_k$ and of molecule 2 by $X_{k+1} + d, \ldots, X_{\mathcal{N}} + d$. Then there exists a distance d_0, depending on the X_i, and rotations $\mathcal{R}_{1,2}$ such that

$$E(\mathcal{R}_1 X_1, \ldots, \mathcal{R}_1 X_k, \mathcal{R}_2 X_{k+1} + d, \ldots, \mathcal{R}_2 X + d)$$

$$\leqslant E(X_1, \ldots, X_k) + E(X_{k+1}, \ldots, X_{\mathcal{N}}) - \frac{c}{d^6}$$

for all $|\mathbf{d}| > d_0$. $E(X_1, \ldots, X_k)$ and $E(X_{k+1}, \ldots, X_{\mathcal{N}})$ are the energies of the isolated molecules where at least one is supposed to be electrically neutral.

Remarks (6.11)

(1) Heuristically the attraction arises because each molecule induces a dipole moment in the other and they orient themselves so as to decrease the energy. It is a typical correlation-effect and cannot be understood in a one-electron picture. To create a dipole moment m_j in molecule j costs and energy $\sim m_j^2$ and what one gains in energy is $-m_1 m_2 / d^3$. However, the minimum over m_j of $c_1 m_1^2 + c_2 m_2^2 - m_1 m_2 / d^3$ for large d is zero and is reached for $m_j = 0$. It turns out that the creation of correlated dipoles costs an energy $cm_1^2 m_2^2$ and $cm_1^2 m_2^2 - m_1 m_2 / d^3$ reaches the minimum $-1/4cd^6$ for $m_1 m_2 = 1/2cd^3$.

(2) The rotation is necessary because the molecules may have permanent multipole moments and one may have to orient them favourably.

(3) The distance d_0 is such that the probability of finding an electron outside this radius is small. Thus one might say that one has van der Waals attraction as long as the clusters do not touch each other.

(4) For several clusters the van der Waals potentials add like scalars and not like dipole potentials. This feature also cannot be understood classically.

Once the atoms start overlapping the situation becomes complex. No simple universal rules seem to emerge and one has to consider case by case. Only in the limit $Z_j \to \infty$ does a unifying picture appear; however, the features important for everyday life vanish.

6.7. The limit of infinite nuclear charge

Shortly after Schrödinger had published his equation Thomas and Fermi had the brilliant idea that one should express the energy by the electron density $\rho(x)$ only. Neglecting the nuclear motion the first terms in (6.4) are the electron kinetic energy ($\equiv K$), the attraction to the nuclei ($\equiv -A$) and the

Coulomb repulsion of the electrons ($\equiv R$). In Thomas–Fermi (TF) they are expressed by ρ as follows:

$$K(\rho) = \frac{3}{5}(3\pi^2)^{2/3} \int d^3x\, \rho^{5/3}(x),$$

$$A(\rho) = \sum_{j=1}^{\mathcal{N}} Z_j \int \frac{d^3x\, \rho(x)}{|x - X_j|}, \tag{7.1}$$

$$R(\rho) = \frac{1}{2} \int \frac{d^3x\, d^3x'}{|x - x'|}\, \rho(x)\rho(x').$$

The ground state energy should be the minimum over all ρ of $E = K - A + R$.

Remarks (7.2)

(1) The expression for K is exact for electrons in a box where $\rho = $ const. For arbitrary wave functions ψ (obeying the Pauli principle) it has been shown (Lieb and Thirring, 1975) that $\langle\psi|K|\psi\rangle \geq c \int d^3x\, \rho^{5/3}(x)$ but only with c smaller than in (7.1). Presumably the inequality holds with the c of K in (7.1) but this has yet to be demonstrated.

(2) Whereas $\langle\psi|A|\psi\rangle = A(\rho)$, in $R(\rho)$ the correlations have been neglected. What could be shown (Lieb and Oxford, 1981) is $\langle\psi|R|\psi\rangle \geq R(\rho) - \gamma \int d^3x\, \rho^{4/3}(x)$.

Also rather late one could demonstrate what the Thomas–Fermi energy

$$E_{\mathrm{TF}}(Z_i, X_i, N) = \min_{\int d^3x \rho(x) = N} (K(\rho) - A(\rho) + R(\rho)) \tag{7.3}$$

has to do with Schrödinger's energy for H_{el} from (6.4)

$$E_{\mathrm{S}}(Z_i, X_i, N) = \min_{\langle\psi|\psi\rangle = 1} \langle\psi|H_{\mathrm{el}}|\psi\rangle. \tag{7.4}$$

The answer is given by (Lieb and Simon, 1977)

Theorem (7.5)

$$E_{\mathrm{TF}}(z_i, X_i, \alpha) = \lim_{Z \to \infty} Z^{-7/3} E_{\mathrm{S}}(Zz_i, Z^{-1/3}X_i, Z\alpha)$$

with $z_i = Z_i/\sum_j Z_j$.

Remarks (7.6)

(1) A well-known argument shows that E_{S} increases with $Z^{7/3}$ and (7.5) tells us that E_{TF} fixes the leading term. The corrections are presumably

$O(Z^{-1/3})$, and for $\mathcal{N}=1$, $R=0$, they can be calculated explicitly. In the general case they are still a matter of dispute.
(2) The scaling $Z^{-1/3}X_i$ tells us that for molecules the Thomas–Fermi limit is also a high density limit.

If we do not fix α, the degree of ionization, and ask for the minimal energy

$$E_{TF}(z_i, X_i) = \min_{\alpha>0} E_{TF}(z_i, X_i, \alpha)$$

we might in (7.3) minimize over $\bar\rho = \lambda\rho$ for any $\lambda>0$. In addition, in the atomic case $(\mathcal{N}=1,\; X_1=0)$ we also can take the minimum over $\bar\rho(x) = \rho(\lambda x)$. Since K, A and R scale differently they are related at the minimum where $\partial E_{TF}/\partial\lambda = 0$ (Thirring, 1983).

Theorem (7.7)
(i) In general $5K - 3A + 6R = 0$.
(ii) For $\mathcal{N}=1$, $X_1=0$, there holds in addition
$$3K - 2A + 5R = 0.$$

Remarks (7.8)
(1) (i) does not imply the virial theorem, we shall see there is no equilibrium position for molecules in TF-theory.
(2) (i) and (ii) imply the virial theorem $2K - A + R = 0$ and fix the ratios of the three contributions to E as $K:A:R = 3:7:1$.

In the atomic case the energy ε depends only on α, $\varepsilon(\alpha) = E_{TF}(1, 0, \alpha)$ corresponding to $\varepsilon(\alpha) = \lim_{Z\to\infty} Z^{-7/3} E_0(Z\alpha, 1/Z)$. If one believes that the better asymptotic form for the atomic energy is $Z^2 N^{1/3}$ one might consider $\bar\varepsilon(\alpha) = \alpha^{-1/3}\varepsilon(\alpha)$ instead. Indeed, the two functions have complementing properties (Lieb, 1981; Thirring, 1983):

Theorem (7.9)
(i) $\varepsilon(\alpha)$, $\bar\varepsilon(\alpha)$ are $\leqslant 0$, $= 0$ iff $\alpha = 0$.
(ii) $\partial\varepsilon/\partial\alpha < 0$ for $0 \leqslant \alpha < 1$, $= 0$ for $\alpha \geqslant 1$, $\partial\bar\varepsilon/\partial\alpha > 0$.
(iii) $\partial^2\varepsilon/\partial\alpha^2 > 0$ for $0 \leqslant \alpha < 1$, $= 0$ for $\alpha > 1$, $\partial^2\bar\varepsilon/\partial\alpha^2 \leqslant \bar\varepsilon'^2/2\bar\varepsilon < 0$.

Remarks (7.10)
(1) (iii) tells us that while ε is convex not only $\bar\varepsilon$ but even $-(-\bar\varepsilon)^{1/2}$ is concave (see (4.13)).
(2) The seemingly contradictory properties (7.9) tell us that the deviation of $-(-\bar\varepsilon)^{1/2}$ from linearity can be at most a few per cent. Since the error of

TF-theory is $Z^{-1/3}$ and therefore at best 20% it seems hardly worth calculating $\varepsilon(\alpha)$ on the computer.
(3) (ii) tells us that there are no negative ions in TF-theory. Keeping Z fixed the energy does not decrease further by increasing N beyond Z.

Passing to the TF-theory of molecules the most striking fact is (Lieb and Simon, 1977)

Theorem (7.11)
With $\alpha = 1$ and

$$\bar{E}_{TF}(z_j, X_j) = E_{TF}(z_j, X_j) + e^2 \sum_{l>k} \frac{z_l z_k}{|X_l - X_k|}$$

we have

$$E_{TF}(z_1 \ldots z_{\mathcal{N}}, X_1 \ldots X_{\mathcal{N}})$$
$$\leqslant E_{TF}(z_1 \ldots z_k, X_1 \ldots X_k) + E_{TF}(z_{k+1} \ldots z_{\mathcal{N}}, X_{k+1} \ldots X_N)$$

but

$$\bar{E}_{TF}(z_1 \ldots z_{\mathcal{N}}, X_1 \ldots X_{\mathcal{N}})$$
$$\geqslant \bar{E}_{TF}(z_1 \ldots z_k, X_1 \ldots X_k) + \bar{E}_{TF}(z_{k+1} \ldots z_{\mathcal{N}}, X_{k+1} \ldots X_{\mathcal{N}}).$$

Remarks (7.12)

(1) Theorem (7.11) shows that in TF-theory the desire of the electrons to have the nuclei close together is never strong enough to overcome the Coulomb repulsion of the nuclei. Thus there is no chemical binding in TF-theory, the binding energy of molecules can never exceed the error of TF-theory (Lieb, 1981).

(2) Iterating (7.11) we see that the energy of any cluster is never below the sum of the energies of its constituent atoms. Thus if all Z_j are below some Z the energy of a cluster is bounded below by $Z^{7/3} \mathcal{N}$.

(3) The most important conclusion one can draw from TF-theory is that matter is stable in Schrödinger's theory. This means that if we have $\mathcal{N}$ atoms and all nuclear charges are bounded by some Z_{max} then the ground state energy is $\geqslant -c \mathcal{N} Z_{max}^{7/3}$. That this is also true in Schrödinger's theory (Dyson and Lenard, 1967; Lieb and Thirring, 1975) follows because TF-theory is not only exact in the limit $Z \to \infty$, $\mathcal{N}$ fixed, but with some corrections (Thirring, 1981b) it also gives a lower bound for the ground state energy in Schrödinger's theory for any Z and $\mathcal{N}$. For arbitrary $\mathcal{N}$ and large Z this bound approaches the TF-theory.

References

Baumgartner, J. (1984) *J. Phys.* A17, 1593

Benguria, R. and Lieb, E. (1983) *Phys. Rev. Lett.* 50, 1771

Combes, J. (1977) Contribution to the Schrödinger equation *Acta Phys. Austr.* 17 (Suppl.)

Daubechies, I. and Lieb, E. (1983) *Commun. Math. Phys.* 90, 511

Dollard, J. (1961) *J. Math. Phys.* 5, 729

Dyson, F. and Lenard, A. (1967) *J. Math. Phys.* 8, 423

Grosse, H. and Pittner, L. (1983) *J. Math. Phys.* 24, 1142

Grosse, H., Narnhofer, H. and Thirring, W. (1979) *J. Phys.* B12, L189

Grosse, H., Grümm, H.-R., Narnhofer, H. and Thirring, W. (1974) *Acta Phys. Austr.* 40, 97

Hertel, P., Lieb, E. and Thirring, W. (1975) *J. Chem. Phys.* 62, 3355

Hill, R. N. (1977) *J. Math. Phys.* 18, 2316

Hoffman-Ostenhof, T. (1980) *J. Phys.* A13, 417

Hunziker, W. (1966) *Helv. Phys. Acta* 39, 451

Kato, T. (1951) *Trans. Am. Math. Soc.* 70, 212

Lieb, E. (1981) *Rev. Mod. Phys.* 53, 603

Lieb, E. (1982) *J. Phys.* A15, 63

Lieb, E. (1984) *Phys. Rev.* A29, 3018

Lieb, E. and Oxford, S. (1981) *Int. J. Quant. Chem.* 19, 427

Lieb, E. and Simon, B. (1977) *Adv. Math.* 23, 22

Lieb, E. and Simon, B. (1978) *J. Phys.* B11, L537

Lieb, E. and Thirring, W. (1975) *Phys. Rev. Lett.* 35, 687

Lieb, E. and Thirring, W. (1986) *Phys. Rev.* A34, 40–6

Lieb, E., Sigal, I. and Simon, B. (1984) *Phys. Rev. Lett.* 52, 994

Moser, J. (1973) *Stable and Random Motions in Dynamical Systems.* Princeton University Press

Narnhofer, H. (1974) *Acta Phys. Austr.* 40, 306

Narnhofer, H. and Thirring, W. (1975) *Acta Phys. Austr.* 41, 281

Narnhofer, H. and Thirring, W. (1980) *Phys. Rev.* 23, 1688

Pearson, D. (1975) *Commun. Math. Phys.* 40, 125

Rebane, H. (1973) *Opt. Spec. (USSR)* 34, 488

Reed, R. and Simon, B. (1972) *Methods of Modern Mathematical Physics*, Vol. I. Academic Press

Reed, R. and Simon, B. (1975) *Methods of Modern Mathematical Physics*, Vol. II. Academic Press

Reed, R. and Simon, B. (1979) *Methods of Modern Mathematical Physics*, Vol. III. Academic Press

Russell, D. and Greenlee (1985) *Phys. Rev. Lett.* 54, 665

Sigal, I. and Soffer, A. (1986) *Bull. Am. Math. Soc.* 14, 108

Thirring, W. (1978) *Classical Dynamical Systems.* Springer

Thirring, W. (1981*a*) *Quantum Mechanics of Atoms and Molecules.* Springer

Thirring, W. (1981*b*) *Commun. Math. Phys.* 79, 1

Thirring, W. (1983) *Quantum Mechanics of Large Systems.* Springer

van Winter, C. (1964) *Mat.-Fys. Danske Vid. Selsk.* 1, 8

Weinstein, A. and Stenger, W. (1972) *Intermediate Problems for Eigenvalues.* Academic Press

Zhislin, G. (1960) *Tr. Mosk. Math. Obs.* 9, 81

7

Molecular dynamics: from H + H₂ to biomolecules

MARTIN KARPLUS

Harvard University

Erwin Schrödinger in his classic volume entitled *What is Life?*, based on a set of lectures delivered in Dublin during 1943, clearly developed the concept that living systems should be understandable by the application of the well-known laws of physics and chemistry. It is only relatively recently, however, that it has become possible (McCammon, Gelin and Karplus, 1977) to study the macromolecules that play an essential role in biology – nucleic acids, proteins and polysaccharides, amongst others – by the theoretical techniques that had been applied previously to small molecules. This new development is providing deeper insights into the structure and dynamics of these macromolecules as a basis for a detailed description of their biological function.

Dynamical simulations for biopolymers are a natural development of the earlier studies of atoms and small molecules. Molecular dynamics has followed two pathways which come together in the study of biomolecular dynamics. One of these, usually referred to as trajectory calculations, has an ancient history that goes back to two-body scattering problems for which analytic solutions can be achieved. However, even for only three particles with realistic interactions, difficulties arise. An example is provided by the simplest chemical reaction, $H + H_2 \rightarrow H_2 + H$, for which a prototype calculation was attempted by Hirschfelder, Eyring and Topley in 1936. They used a semi-empirical potential for the interactions of the three hydrogen atoms that was based on the Heitler–London description of the hydrogen molecule, one of the triumphs of Schrödinger's wave mechanics. Hirschfelder and his colleagues were able to calculate a few steps along one trajectory. It was nearly 30 years later that the availability of computers made it possible to complete the calculation. Much has been done since then in applying classical trajectory methods to a wide range of chemical reaction (Porter, 1974; Walker and Light, 1980). These classical studies have been supplemented by semi-classical and quantum mechanical calculations in areas where quantum effects can play an important role (Schatz and Kuppermann, 1980; Walker and Light, 1980). The focus of trajectory studies is now on more complex molecules, their redistribution of internal energy, and the role of this on their reactivity.

The other pathway in molecular dynamics has been concerned with physical rather than chemical interactions (in analogy of physisorption versus chemisorption) and the thermodynamic and average dynamic properties of a large number of particles at equilibrium, rather than detailed trajectories of a few particles. Although the basic ideas go back to van der Waals and Boltzmann, the modern area began with the work of Alder and Wainwright on hard sphere liquids in the late 1950s (Alder and Wainwright, 1959). The paper by Rahman (1964), on a molecular dynamics simulation of liquid argon with a soft sphere (Lennard–Jones) potential represented an important step. Simulations of complex liquids followed; the now classic study of liquid water by Stillinger and Rahman was published in 1974 (Stillinger and Rahman, 1974). Since then, there have been many studies on the equilibrium and nonequilibrium behavior of a wide range of systems (Hoover, 1983; Wood and Erpenbeck, 1976).

This background set the stage for the development of molecular dynamics of biomolecules. The size of an individual molecule, composed of 500 or more atoms for even a small protein, is such that its simulation in isolation can determine approximate equilibrium properties, as in the molecular dynamics of fluids, though detailed aspects of the atomic motions are of considerable interest, as in trajectory calculations. A basic assumption in initiating such studies was that potential functions could be constructed which were sufficiently accurate to give meaningful results for systems as complex as proteins or nucleic acids. In addition, it was necessary to assume that for such inhomogeneous systems, in contrast to the homogeneous character of even 'complex' liquids like water, simulations of an attainable time scale (10 ps in the initial studies) could provide a useful sample of the phase space in the neighborhood of the native structure. For neither of these assumptions was there strong supporting evidence. Nevertheless, it seemed worthwhile in 1975 to apply the techniques of molecular dynamics with the available potential functions to the internal atomic motions of proteins with known crystal structure (McCammon, Gelin and Karplus, 1977; for early reviews of experimental and theoretical developments see Gurd and Rothgeb, 1979; and Karplus and McCammon, 1981).

The most important consequence of the first simulations of biomolecules was that they introduced a conceptual change. Although to chemists and physicists it is self-evident that polymers like proteins and nucleic acids undergo significant fluctuations at room temperature, the classic view of such molecules in their native state had been static in character. This followed from the dominant role of high-resolution X-ray crystallography in providing structural information for these complex systems. The remarkable detail evident in crystal structures led to an image of biomolecules with every atom fixed in place. D.C. Phillips, who determined the first enzyme crystal structure, wrote recently 'the period 1965–75 may be described as the decade of the rigid macromolecule. Brass models of DNA and a variety of proteins dominated the scene and much of the thinking' (Phillips, 1981). Molecular dynamics simulations have been instrumental in changing the static view of the structure of biomolecules to a dynamic picture. It is now recognized that the atoms of which biopolymers are composed are in a state of constant motion at ordinary temperatures. The X-ray structure of a protein provides the average atomic positions, but the atoms exhibit fluid-like motions of sizeable amplitudes about these averages. Crystallographers have acceded to this viewpoint and have come so far as to sometimes emphasize the parts of a molecule they do not see in a crystal structure as evidence of motion or disorder (Marquart *et al.*, 1980). The new understanding of protein dynamics subsumes the static picture in that use of the average positions still allows discussion of many aspects of biomolecule function in the language of structural chemistry. However, the recognition of the importance of fluctuations opens the way for more sophisticated and accurate interpretations.

Simulation studies in this area, as in others, have the possibility of providing the ultimate detail concerning motional phenomena. The primary limitation of simulation methods is that they are approximate. It is here that experiment plays an essential role in validating the simulation methods; that is, comparisons with experimental data can serve to test the accuracy of the calculated results and to provide criteria for improving the methodology. When experimental comparisons indicate that the simulations are meaningful, their capacity for providing detailed results often makes it possible to examine specific aspects of the atomic motions far more easily than by making measurements.

At the present stage of the molecular dynamics of biomolecules, there is a general understanding of the motion that occurs on a subnanosecond time scale; that is, the types of motion have been demonstrated, their characteristics evaluated and the important factors determining their properties delineated. Simulation methods have shown that the structural fluctuations are sizeable; particularly large fluctuations are found where steric constraints due to molecular packing are small (e.g. in the exposed sidechains and external loops), but substantial mobility is also found in the interior of a macromolecule. Local atomic displacements in the interior are correlated in a manner that tends to minimize disturbances of the global structure. This leads to fluctuations larger than would be permitted in a rigid protein matrix.

For motions on a longer time scale, our understanding is more limited. When the motion of interest can be described in terms of a reaction path (e.g. hinge-bending, local activated events), methods exist for examining the nature and the rate of the process. However, for the motions that are slow due to their complexity and involve large-scale structural changes, extensions of the available approaches are required. Harmonic and simplified model dynamics, as well as reaction-path calculations, can provide information on slower motions, such as opening fluctuations and helix–coil transitions.

In applying molecular dynamics to physical studies of biomolecules (such as X-ray, nuclear magnetic resonance, infra-red, Raman, inelastic neutron scattering, fluorescence depolarization, and so on), a number of aspects are important. There is of course the direct comparison between the results of calculations and the experimental data. More interesting is the possibility of extending the interpretation of experiments. Also, experimental data can be generated by simulations and analyzed as would real data to test the method used. Finally, new effects can be predicted from the simulation as a stimulus for additional experimental investigations.

In what follows, applications of molecular dynamics that illustrate each of these points are outlined.

7.1. A test: X-ray diffraction

Since atomic fluctuations are the basis of protein dynamics, it is important to have experimental tests of the accuracy of the simulation results concerning them. For the magnitudes of the motions, the most detailed data are provided, in principle, by an analysis of the Debye–Waller or temperature factors obtained in crystallographic refinements of X-ray structures.

It is well known from small molecule crystallography that the effects of thermal motion must be included in the interpretation of the X-ray data to obtain accurate structural results. Detailed models have been introduced to take account of anisotropic and anharmonic motions of the atoms and these models have been applied to high resolution data for small molecules (Zucker and Schulz, 1982). In protein crystallography, the limited data available, relative to the large number of parameters that have to be determined, have made it necessary to assume that the atomic motions are isotropic and harmonic. In that case the structure factor, $F(\mathbf{Q})$, which is related to the measured intensity by $I(\mathbf{Q}) = |F(\mathbf{Q})|^2$, is given by

$$F(\mathbf{Q}) = \sum_{j=1}^{N} f_j(\mathbf{Q}) \, e^{i\mathbf{Q} \cdot \langle \mathbf{r}_j \rangle} \, e^{W_j(\mathbf{Q})}, \qquad (1.1)$$

where $\mathbf{Q}$ is the scattering vector, $\langle \mathbf{r}_j \rangle$ is the average position of atom j with atomic scattering factor $f_j(\mathbf{Q})$ and the sum is over the N atoms in the asymmetric unit of the crystal. The Debye–Waller factor, $W_j(\mathbf{Q})$, is defined by

$$W_j(\mathbf{Q}) = -\tfrac{8}{3}\pi^2 \langle \Delta r_j^2 \rangle s^2 = -B_j s^2, \qquad (1.2)$$

where $s = |\mathbf{Q}|/4\pi$. The quantity B_j is usually referred to as the temperature factor, which is directly related to the mean-square atomic fluctuations in the isotropic harmonic model. More generally, if the motion is harmonic but anisotropic, a set of six parameters

$$B_j^{xx} = \langle \Delta x_j^2 \rangle, \qquad B_j^{xy} = \langle \Delta x_j \, \Delta y_j \rangle, \ldots, B_j^{zz} = \langle \Delta z_j^2 \rangle$$

is required to fully characterize the atomic motion. Although in the earlier X-ray studies of proteins, the significance of the temperature factors was ignored (presumably because the data were not at a sufficient level of resolution and accuracy), more recently attempts have been made to relate the observed temperature factors to the atomic motions. In principle, the temperature factors provide a very detailed measure of the motions because information is available for the mean-square fluctuation of each heavy atom. In practice, there are two types of difficulties in relating the B factors obtained from protein refinements to the atomic motions. The first is that, in addition to thermal fluctuations, any static (lattice) disorder in the crystal contributes to the B factors; i.e., since a crystal is made up of many unit cells, different molecular geometries in the various cells have the same effect on the average electron density, and therefore the B factor, as atomic motions. In only one case, the iron atom of myoglobin, has there been an experimental attempt to determine the disorder contribution (Hartmann *et al.*, 1982). Since the Mossbauer effect is not altered by static disorder (i.e. each nucleus absorbs independently) but does depend on atomic motions, comparisons of Mossbauer and X-ray data have been used to estimate a disorder contribution for the iron atom; the value obtained is

$$\langle \Delta r_{\text{Fe}}^2 \rangle = 0.08 \text{ Å}^2.$$

Although the value is only approximate, it nevertheless indicates that the observed B factors (e.g. of the order of 0.44 Å^2 for backbone atoms and 0.50 Å^2 for sidechain atoms) are dominated by the motional contribution. Most experimental B factor values are compared directly with the molecular dynamics results (i.e. neglecting the disorder contribution) or are rescaled by a constant amount (e.g. by setting the smallest observed B factor to zero) on the assumption that the disorder contribution is the same for all atoms (Petsko and Ringe, 1984).

Second, since simulations have shown that the atomic fluctuations are highly anisotropic and, in some cases, anharmonic, it is important to determine the errors introduced into the refinement process by the assumption of isotropic and harmonic motion. A direct experimental estimate of the errors is difficult because sufficient data are not yet available for protein crytals. Moreover, any data set includes other errors which would obscure the analysis. As an alternative to an experimental analysis of the errors in the refinement of proteins, a purely theoretical approach can be used (Kuriyan *et al.*, 1986). The basic idea is to generate X-ray data from a molecular dynamics simulation of a protein and to use these data in a standard refinement procedure. The error in the analysis can then be determined by comparing the refined X-ray structure and temperature factors with the average structure and the mean-square fluctuations from the simulation. Such a comparison, in which no real experimental results are used, avoids problems due to inaccuracies in the measured data (exact calculated intensities are used), to crystal disorder (there is none in the model), and to approximations in the simulation (the simulation is exact for this case). The only question about such a comparison is whether the atomic motions found in the simulation are a meaningful representation of those occurring in proteins. As has been shown (Karplus and McCammon, 1983; Petsko and Ringe, 1984; Porter, 1974), molecular dynamics simulations provide a reasonable picture of the motions in spite of errors in the potentials, the neglect of the crystal environment and the finite time classical trajectories used to obtain the results. However, these inaccuracies do not affect the exactitude of the computer 'experiment' for testing the refinement procedure that is described below.

In this study (Kuriyan *et al.*, 1986), a 25 ps molecular dynamics trajectory for myoglobin was used (Levy *et al.*, 1985). The average structure and the mean-square fluctuations from the structure were calculated directly from the trajectory. To obtain the average electron density, appropriate atomic electron distributions were assigned to the individual atoms, and the results for each coordinate set were averaged over the trajectory. Given the symmetry, unit cell dimensions and position of the myoglobin molecule in the unit cell, average structure factors, $\langle F(\mathbf{Q}) \rangle$, and intensities, $I(\mathbf{Q}) = |\langle F(\mathbf{Q}) \rangle|^2$, were calculated from the Fourier transform of the average electron density, $\langle \rho(\mathbf{r}) \rangle$, as a function of position $\mathbf{r}$ in the unit cell. Data were generated at 1.5 Å resolution, as this is comparable to the resolution of the best X-ray data currently available for proteins the size of myoglobin (Kuriyan, Petsko and Karplus, unpublished; Phillips, 1980). The resulting intensities at Bragg reciprocal lattice points were used as input data for the widely applied crystallographic program, PROLSQ (Konnert and Hendrickson, 1980). The time-averaged atomic positions obtained from the simulation and a uniform temperature factor provide the initial model for refinement. The positions and an isotropic, harmonic temperature factor for each atom were then refined iteratively against the computer generated intensities in the standard way. Differences between the refined results for the average atomic positions and their mean-square fluctuations and those obtained from the molecular dynamics trajectories are due to errors introduced by the refinement procedure.

The overall root-mean-square (rms) error in atomic positions ranged from 0.24 to 0.29 Å for slightly different restrained and unrestrained refinement procedures (Kuriyan *et al.*, 1986). The errors in backbone positions (0.10–0.20 Å) are generally less than those for sidechain atoms (0.28–0.33 Å); the largest positional errors are of the order of 0.6 Å. The backbone errors, though small, are comparable to the rms deviation of 0.21 Å between the positions of the backbone atoms in the refined experimental structures of oxymyoglobin and carboxy myoglobin (Kuriyan, Petsko and Karplus, unpublished; Phillips, 1980). Further, the positional errors are not uniform over the whole structure. There is a strong correlation between the positional error and the magnitude of the mean-square fluctuation for an atom, with certain regions of the protein, such as loops and external sidechains, having the largest errors.

The refined mean-square fluctuations are systematically smaller than the fluctuations calculated directly from the simulation. The magnitude and variation of temperature factors along the backbone are relatively well reproduced, but the refined sidechain fluctuations are almost always significantly smaller than the actual values. The average backbone B factors from different refinements are in the range 11.3–11.7 Å^2, as compared with the exact value of 12.4 Å; for the sidechains, the refinements yield 16.5–17.6 Å^2, relative to the exact value of 26.8 Å^2. Regions of the protein that have high mobility have large errors in temperature factors as well as in positions. Examination of all atoms shows that fluctuations greater than about 0.75 Å^2 ($B=20$ Å^2) are almost always underestimated by the refinement. Moreover, while actual mean-square atomic fluctuations have values as large as 5 Å^2, the X-ray refinement leads to an effective upper limit of about 2 Å^2. This arises from the fact that most of the atoms with large fluctuations have multiple conformations and that the refinement procedure picks out one of them.

To do refinements that take some account of anisotropic motions for all but the smallest proteins, it has been necessary to introduce assumptions concerning the nature of the anisotropy. One possibility is to assume anisotropic rigid body motions for sidechains such as tryptophan and phenylalanine (Artymix *et al.*, 1979; Glover *et al.*, 1983). An alternative is to introduce a 'dictionary' in which the orientation of the anisotropy tensor is related to the stereochemistry around each atom (Konnert and Hendrickson, 1980); this reduces the six independent parameters of the anisotropic temperature factor tensor B_j to three parameters per atom. An analysis of a simulation for BPTI (Yu, Karplus and Hendrickson, 1985) has shown that the actual anisotropies in the atomic motions are generally not simply related to the local stereochemistry; an exception is the mainchain carbonyl oxygen which has its largest motion perpendicular to the C$=$O bond. Thus, use of stereochemical assumptions in the refinement can yield incorrectly oriented anisotropy tensors and significantly reduced values for the anisotropies. The large scale motions of atoms are collective, and sidechains tend to move as a unit so that the directions of largest motion are not related to the local bond direction, and have similar orientations in the different atoms forming a group that is undergoing correlated motions. This means that it is necessary to use the full anisotropy tensor to obtain meaningful results. This is possible with proteins that are particularly well ordered so that the diffraction data extend to better than 1 Å resolution.

7.2. An extension: nuclear magnetic resonance

Nuclear magnetic resonance (NMR) is an experimental technique that has played an essential role in the analysis of the internal motions of proteins (Campbell, Dobson and Williams, 1978; for early reviews of experimental and theoretical developments see Gurd and Rothgeb, 1979; and Karplus and McCammon, 1981). Like X-ray diffraction, it can provide information about individual atoms; unlike X-ray diffraction, NMR is sensitive not only to the magnitude but also to the time scales of the motions. Nuclear relaxation processes are dependent on atomic motions on the nanosecond to picosecond time scale. Although molecular tumbling is generally the dominant relaxation mechanism for proteins in solution, internal motions contribute as well; for solids, the internal motions are of primary importance. In addition, NMR parameters, such as nuclear spin–spin coupling constants and chemical shifts, depend on the protein environment. In many cases different local conformations exist but the interconversion is rapid on the NMR time scale, here on the order of milliseconds, so that average values are observed. When the interconversion time is on the order of the NMR time scale or slower, the transition rates can be studied; an example is provided by the reorientation of aromatic rings (Campbell *et al.*, 1976; Wagner, DeMarco and Wuthrich, 1976).

In addition to supplying data on the dynamics of proteins, NMR can also be used to obtain structural information. With recent advances in techniques it is now possible to obtain a large number of approximate interproton distances for proteins by the use of nuclear Overhauser effect (NOE) measurements (Noggle and Schirmer, 1971). If the protein is relatively small and has a well-resolved spectrum, a large portion of the protons can be assigned, and several hundred distances for these protons can be determined by the use of two-dimensional NMR techniques (Wagner and Wuthrich, 1982). Clearly these distances can serve to provide structural information for proteins, analogous to their earlier use for organic molecules (Honig *et al.*, 1971; Noggle and Schirmer, 1971). Of great interest is the possibility that enough distance information can be measured to actually determine the high resolution structure of a protein in solution, to supplement results from crystallography, particularly for proteins that are difficult to crystallize. In what follows we consider two questions related to this possibility. The first concerns the effect of motional averaging on the accuracy of the apparent distances obtained from the NOE studies and the second, whether the number of distances that can be obtained experimentally are sufficient for a structure determination.

For spin-lattice relaxation, such as observed in NOE measurements, it is possible to express the behavior of the magnetization of the nuclei being studied by the equation (Olejniczak *et al.*, 1984; Solomon, 1955)

$$\frac{d(I_z(t) - I_0)_i}{dt} = -\rho_i (I_z(t) - I_0)_i - \sum_{i \neq j} \sigma_{ij} (I_z(t) - I_0)_j, \qquad (2.1)$$

where $I_z(t)_i$ and I_{0i} are the z components of the magnetization of nucleus i, ρ_i is the direct relaxation rate of nucleus i, and σ_{ij} is the cross relaxation rate between nuclei i and j. The quantities ρ_i and σ_{ij} can be expressed in terms of spectral densities

$$\rho_i = \frac{6\pi}{5} \gamma_i^2 \gamma_j^2 \hbar^2 \sum_{i \neq j} \left[\tfrac{1}{3} J_{ij}(\omega_i - \omega_j) + J_{ij}(\omega_i) + 2 J_{ij}(\omega_i + \omega_j) \right] \qquad (2.2)$$

$$\sigma_{ij} = \frac{6\pi}{5} \gamma_i^2 \gamma_j^2 \hbar^2 \left[2 J_{ij}(\omega_i + \omega) - \tfrac{1}{3} J_{ij}(\omega_i - \omega_j) \right]. \qquad (2.3)$$

The spectral density functions can be obtained from the correlation functions for the relative motions of the nuclei with spins i and j (Levy, Karplus and Wolynes, 1981; Olejniczak *et al.*, 1984):

$$J_{ij}^n(\omega) = \int_0^\infty \frac{\langle Y_n^2(\Theta_{\text{lab}}(t)\phi_{\text{lab}}(t)) Y_n^{2*}(\Theta_{\text{lab}}(0)\phi_{\text{lab}}(0)) \rangle}{r_{ij}^3(0) r_{ij}^3(t)} \cos(\omega t)\, \mathrm{d}t \qquad (2.4)$$

where $Y_n^2(\Theta(t)\phi(t))$ are second-order spherical harmonics and the angular brackets represent an ensemble average which is approximated by an integral over the molecular dynamics trajectory. The quantities $\Theta_{\text{lab}}(t)$ and $\phi_{\text{lab}}(t)$ are the polar angles at time t of the internuclear vector between protons i and j with respect to the external magnetic field and r_{ij} is the interproton distance. In the simplest case of a rigid molecule undergoing isotropic tumbling with a correlation time τ_0 this reduces to the familiar expression

$$J_{ij}(\omega) = \frac{1}{4\pi r_{ij}^6} \left[\frac{\tau_0}{1 + (\omega \tau_0)^2} \right] \qquad (2.5)$$

The NOE corresponds to the selective enhancement of a given resonance in a dipolar coupled spin system. Of particular interest for obtaining motional and distance information are measurements that provide time dependent NOEs from which the cross relaxation rates σ_{ij} (2.3) can be determined directly or indirectly by solving a set of coupled equations (2.1)–(2.3). Motions on the picosecond time scale are expected to introduce averaging effects that decrease the cross relaxation rates by a scale factor relative to the rigid model. A lysozyme molecular dynamics simulation (Ichiye *et al.*, 1986) has been used to calculate dipole vector correlation functions (Olejniczak *et al.*, 1984) for proton pairs that have been studied experimentally (Olejniczak, Poulsen and Dobson, 1981; Poulsen, Hoch and Dobson, 1980). Four proton pairs on three sidechains (Trp 28, Ile 98 and Met 105) with very different motional properties were examined. Trp 28 is quite rigid, Ile 98 has significant fluctuations, and Met 105 is particularly mobile in that it jumps among different sidechain conformations during the simulation. The rank order of the scale factors (order parameters) is the same in the theoretical and experimental results. However, although the results for the Trp 28 protons agree with the measurements to within the experimental error, for both Ile 98 and Met 105 the motional averaging found from the NOEs is significantly greater than the calculated value. This suggests that these residues are undergoing rare fluctuations involving transitions that are not adequately sampled by the simulation.

If NOEs are measured between pairs of protons whose distance is not fixed by the structure of a residue, the strong distance dependence of the cross relaxation rates ($1/r^6$) can be used to obtain estimates of the interproton distances (Clore *et al.*, 1985; Olejniczak *et al.*, 1981; Poulsen *et al.*, 1980; Wagner and Wuthrich, 1982). The simplest application of this approach is to assume that proteins are rigid and tumble isotropically. The lysozyme molecular dynamics simulation was used to determine whether picosecond fluctuations are likely to introduce important errors into such an analysis (Olejniczak *et al.*, 1984). The results show that the presence of the motions will cause a general decrease in most NOE effects observed in a protein. However, because the distance depends on the sixth root of the observed NOE, motional errors of a factor of two in the latter lead to only a 12 % uncertainty in the distance. Thus, the decrease is usually too small to produce a significant change in the distance estimated from the measured NOE value. This is consistent with the excellent correlation found between experimental NOE values and those calculated using distances from a crystal structure (Poulsen *et al.*, 1980). Specific NOEs can, however, be altered by the internal motions to such a degree that the effective distances obtained are considerably different from those predicted for a static structure. Such possibilities must, therefore, be considered in any structure determination based on NOE data. This is true particularly for cases involving averaging over large scale fluctuations.

Because of the inverse sixth power of the NOE distance dependence, experimental data so far are limited to protons that are separated by less than 5 Å. Thus, the long-range information required for a direct protein structure determination is not available. To overcome this limitation it is possible to introduce additional information provided by empirical energy functions (Brooks *et al.*, 1983). One way of proceeding is to do molecular dynamics simulations with the approximate interproton distances introduced as restraints in the form of skewed biharmonic potentials (Brunger *et al.*, 1986; Clore *et al.*, 1985) with the force constants chosen to correspond to the experimental uncertainty in the distance.

A model study of the small protein crambin (Brunger *et al.*, 1986) was made with realistic NOE restraints; 250 approximate interproton distances less than 4 Å were used, including 184 short-range distances (i.e. connecting protons in two residues that were less than five residues apart in the sequence) and 56 long-range distances. The molecular dynamics simulations converged to the known crambin structure from different initial extended structures. The average structure obtained from the simulations with a series of different protocols had rms deviations of 1.3 Å for the backbone atoms, and 1.9 Å for the sidechain atoms. Individual converged simulations had rms deviations in the range 1.5–2.1 Å and 2.1–2.8 Å for the backbone and sidechain atoms, respectively. Further, it was shown that a dynamics structure with significantly larger deviations (5.7 Å) could be characterized as incorrect, independent of a knowledge of the crystal structure because of its higher energy and the fact that the NOE restraints were not satisfied within the limits of error. The incorrect structure resulted when all NOE restraints were introduced simultaneously, rather than allowing the dynamics to proceed first in the presence of only the short-range restraints followed by introduction of the long-range restraints. Also of interest is the fact that although crambin has three disulfide bridges it was not necessary to introduce information concerning them to obtain an accurate structure.

The folding process as simulated by the restrained dynamics is very rapid. At the end of the first 2 ps the secondary structure is essentially established while the molecule is still in an extended conformation. Some tertiary folding occurs even in the absence of long-range restraints. When they are introduced, it takes about 5 ps to obtain a tertiary structure that is approximately correct and another 6 ps to introduce the small adjustments required to converge to the final structure.

It is of interest to consider whether the results obtained in the restrained dynamics simulation have any relation to actual protein folding. That correctly folded structures are achieved only when the secondary structural elements are at least partly formed before the tertiary restraints are introduced is suggestive of the diffusion–collision model of protein folding (Bashford, Weaver and Karplus, 1984). Clearly, the specific pathway has no physical meaning since it is dominated by the NOE restraints. Also, the time scale of the simulated folding process is 12 orders of magnitude faster than experimental estimates. About six to nine orders of magnitude of the rate increase are due to the fact that the secondary structure is stable once it is formed, in contrast to a real protein where the secondary structural elements spend only a small fraction of time in the native conformation until coalescence has occurred. The remainder of the artificial rate increase presumably arises from the fact that the protein follows a single direct path to the folded state in the presence of the NOE restraints, instead of having to go through a complex search process.

7.3. A prediction: structural role of active site waters in ribonuclease A

To achieve a realistic treatment of the solvent-accessible active sites, a new molecular dynamics simulation method, called the stochastic boundary method, has been implemented (Brooks and Karplus, 1983; Brooks, Brunger and Karplus, 1985; Brunger, Brooks and Karplus, 1984). It makes possible the simulation of a localized region, approximately spherical in shape, that is composed of the active site with or without ligands, the essential portions of the protein in the neighborhood of the active site, and the surrounding solvent. The approach provides a simple and convenient method for reducing the total number of atoms included in the simulation, while avoiding spurious edge effects.

The stochastic boundary method for solvated proteins starts with a known X-ray structure; for the present problem the refined high resolution (1.5–2 Å) X-ray structures provided by Petsko and co-workers was used (Campbell and Petsko, 1986; Gilbert, Fink and Petsko, 1986). The region of interest (here the active site of ribonuclease A) was defined by choosing a reference point (which was taken at the position of the phosphorus atom in the CpA inhibitor complex) and constructing a sphere of 12 Å radius around this point. Space within the sphere not occupied by crystallographically determined atoms was filled by water molecules, introduced from an equilibrated sample of liquid water. The 12 Å sphere was further subdivided into a reaction region (10 Å radius) treated by full molecular dynamics and a buffer region (the volume between 10 and 12 Å) treated by Langevin dynamics, in which Newton's equations of motion for the non-hydrogen atoms are augmented by a fractional term and a random-force term; these additional terms approximate the effects of the neglected parts of the system and permit energy transfer in and out of the reaction region. Water molecules diffuse freely between the reaction and buffer regions but are prevented from escaping by an average boundary force (Brunger *et al.*, 1984). The protein atoms in the buffer region are constrained by harmonic forces derived from crystallographic temperature factors (Brooks *et al.*, 1985). The forces on the atoms and their dynamics were calculated with the CHARMM program (Brooks *et al.*, 1983); the water molecules were represented by the ST2 model (Stillenger and Rahman, 1974).

One of the striking aspects of the active site of ribonuclease is the presence of a large number of positively charged groups, some of which may be involved in guiding and/or binding the substrate (Matthew and

Richards, 1982). The simulation demonstrated that these residues are stabilized in the absence of ligands by well-defined water networks. A particular example includes Lys-7, Lys-41, Lys-66, Arg-39 and the doubly protonated His-119. Bridging waters, some of which are organized into trigonal bipyramidal structures, were found to stabilize the otherwise very unfavorable configuration of near-neighbor positive groups because the interaction energy between water and the charged $C-NH_n^+$ ($n = 1, 2$ or 3) moieties is very large; e.g., at a donor-acceptor distance of 2.8 Å, the $C-NH_3^+-H_2O$ energy is -19 kcal/mol with the empirical potential used for the simulation (Brooks *et al.*, 1983), in approximate agreement with accurate quantum mechanical calculations (Desmeules and Allen, 1980) and gas-phase ion-molecules data (Kebarle, 1977). The average stabilization energy of the charged groups (Lys-7, Lys-41, Lys-66, Arg-39 and His-119) and the 106 water molecules included in the simulation is -376.6 kcal/mol. This energy is calculated as the difference between the simulated system and a system composed of separate protein and bulk water. Unfavorable protein–protein charged-group interactions are balanced by favorable water–protein and water–water interactions. The average energy per molecule of pure water from an equivalent stochastic boundary simulation (Brunger *et al.*, 1984) was -9.0 kcal/mol, whereas that of the waters included in the active site simulation was -10.2 kcal/mol; in the latter a large contribution to the energy came from the interactions between the water molecules and the protein atoms. It is such energy differences that are essential to a correct evaluation of binding equilibria and the changes introduced by site-specific mutagenesis (Fersht *et al.*, 1985).

During the simulation, the water molecules involved in the charged-group interactions oscillated around their average positions, generally without performing exchange. On a longer time scale, it is expected that the waters would exchange and that the sidechains would undergo larger scale displacements. This is in accord with the disorder found in the X-ray results for lysine and arginine residues (e.g. Lys-41 and Arg-39) (Gilbert *et al.*, 1986; Wlodawer, 1985), a fact that makes difficult a crystallographic determination of the water structure in this case. It is also of interest that Lys-7 and Lys-41 have an average separation of only 4 Å in the simulation, less than that found in the X-ray structure. That this like charged pair can exist in such a configuration is corroborated by experiments that have shown that the two lysines can be crosslinked (Marfey, Uziel and Little, 1965); the structure of this compound has been reported recently (Weber *et al.*, 1985) and is similar to that found in the native protein.

In addition to the role of water in stabilizing the charged groups that span the active site and participate in catalysis, water molecules make hydrogen bonds to protein polar groups that become involved in ligand binding. A particularly clear example is provided by the adenine-binding site in the CpA simulation. The NH_2 group of adenine acted as a donor, making hydrogen bonds to the carbonyl of Asn-67, and the ring N^{1A} acted as an acceptor for a hydrogen bond from the amide group of Glu-69. Corresponding hydrogen bonds were present in the free ribonuclease simulation, with appropriately bound water molecules replacing the substrate. These waters and those that interact with the pyrimidine-site residues Thr-45 and Ser-123 help to preserve the protein structure in the optimal arrangement for binding. Similar substrate 'mimicry' has been observed in X-ray structures of lysozyme (Blake, Pulford and Artymiuk, 1983) and of penicillopepsin (James and Sielecki, 1983), but has not yet been seen in ribonuclease.

Acknowledgement

I gratefully acknowledge the work done by many collaborators in the research described in this review. Their essential contributions are made clear by the citations in the reference list.

References

Alder, B. J. and Wainwright, J. (1959) *J. Chem. Phys.* 31, 459

Artymiuk, P. J., Blake, C. C. F., Grace, D. E. P., Oatley, S. J., Phillips, D. C. and Sternberg, J. J. E. (1979) *Nature* 280, 563

Bashford, D., Weaver, D. L. and Karplus, M. (1984) *J. Biol. Struct. Dyn.* 1, 1243

Blake, C. C. F., Pulford, W. C. A. and Artymiuk, P. J. (1983) *J. Mol. Biol.* 167, 693

Brooks, B. R., Bruccoleri, R. E., Olafson, B. D., States, D. J., Swaminathan, S. and Karplus, M. (1983) *J. Comp. Chem.* 4, 187

Brooks, C. L. and Karplus, M. (1983) *J. Chem. Phys.* 79, 6312

Brooks, C. L., Brunger, A. and Karplus, M. (1985) *Biopolymers* 24, 843

Brunger, A. T., Brooks, C. L. and Karplus, M. (1984) *Chem. Phys. Lett.* 105, 495

Brunger, A. T., Clore, G. M., Gronenborn, A. M. and Karplus, M. (1986) *Proc. Natl. Acad. Sci. USA* 83, 380

Campbell, I. D., Dobson, C. M. and Williams, R. J. P. (1978) *Adv. Chem. Phys.* 39, 55

Campbell, I. D., Dobson, C. M., Moore, G. R., Perkins, S. J. and Williams, R. J. P. (1976) *FEBS Lett.* 70, 96

Campbell, R. L. and Petsko, G. A. (1986) *Biochemistry* (in press)

Clore, G. M., Gronenborn, A. M., Brunger, A. T. and Karplus, M. (1985) J. Mol. Biol. 186, 435

Desmeules, D. J. and Allen, L. C. (1980) J. Chem. Phys. 72, 4731

Fersht, A. R., Shi, J-P., Knill-Jones, J., Lowe, D. M., Wilkinson, A. J., Blowq, D. M., Brick, P., Carter, P., Waye, M. M. Y. and Winter, G. (1985) Nature 314, 235

Gilbert,W. A., Fink, A. L. and Petsko, G. A. (1986) Biochemistry (in press)

Glover, I., Haneef, I., Pitts, J., Wood, S., Moss, D., Tickle, I. and Blundell, T. (1983) Biopolymers 22, 293

Gurd, F. R. N. and Rothgeb, J. M. (1979) Adv. Prot. Chem. 33, 73

Hartmann, H., Parak, F., Steigemann, W., Petsko, G. A., Ponzi, D. R. and Frauenfelder, H. (1982) Proc. Natl. Acad. Sci. USA 79, 4967

Honig, B., Hudson, B., Sykes, B. D. and Karplus, M. (1971) Proc. Natl. Acad. Sci. USA 68, 1289

Hoover, W. G. (1983) Ann. Rev. Phys. Chem. 34, 103

Ichiye, T., Olafson, B., Swaminathan, S. and Karplus, M. (1986) Biopolymers (in press)

James, M. N. G. and Sielecki, A. R. (1983) J. Mol. Biol. 163, 299

Karplus, M. and McCammon, J. A. (1981) CRC Crit. Rev. Biochem. 9, 293

Karplus, M. and McCammon, J. A. (1983) Ann. Rev. Biochem. 52, 263

Kebarle, P. (1977) Ann. Rev. Phys. Chem. 28, 445

Konnert, J. H. and Hendrickson, W. A. (1980) Acta Cryst. A36, 344

Kuriyan, J., Petsko, G. A., Levy, R. M. and Karplus, M. (1986) J. Mol. Biol. (in press)

Levy, R. M., Karplus, M. and Wolynes, P. G. (1981) J. Am. Chem. Soc. 103 5998

Levy, R. M., Sheridan, R. P., Keepers, J. W., Dubey, G. S., Swaminathan, S. and Karplus, M. (1985) Biophys. J. 48, 509

Marfey, P. S., Uziel, M. and Little, J. (1965) J. Biol. Chem. 240, 3270

Marquart, M., Deisendorfer, J., Huber, R. and Palm, W. (1980) J. Mol. Biol. 141, 369

Matthew, J. B. and Richards, F. M. (1982) Biochemistry 21, 4989

McCammon, J. A., Gelin, B. R. and Karplus, M. (1977) Nature 267, 585

Noggle, J. H. and Schirmer, R. E. (1971) The Nuclear Overhauser Effect. New York: Academic Press

Olejniczak, E. T., Poulsen, F. M. and Dobson, D. M. (1981) J. Am. Chem. Soc.103, 6574

Olejniczak, E. T., Dobson, C. M., Karplus, M. and Levy, R. M. (1984) J. Am. Chem. Soc. 106, 1923

Petsko, G. A. and Ringe, D. (1984) Ann. Rev. Biophys. Bioeng. 13, 331

Phillips, D. C. (1981) in Biomolecular Stereodynamics (Sarma, R. H., ed.), p. 497. New York: Adenine

Phillips, S. E. V. (1980) J. Mol. Biol. 142, 531

Porter, R. N. (1974) Ann. Rev. Phys. Chem. 25, 371

Poulsen, F. M., Hoch, J. C. and Dobson, C. M. (1980) Biochemistry 19, 2597

Rahman, A. (1964) Phys. Rev. A136, 405

Schatz, G. C. and Kuppermann, A. (1980) J. Chem. Phys. 62, 2502

Solomon, I. (1955) *Phys. Rev.* 99, 559

Stillinger, F. H. and Rahman, A. (1974) *J. Chem. Phys.* 60, 1545

Wagner, G. and Wuthrich, K. (1982) *J. Mol. Biol.* 160, 343

Wagner, G., DeMarco, A. and Wuthrich, K. (1976) *Biophys. Struct. Mech.* 2, 139

Walker, R. B. and Light, J. C. (1980) *Ann. Rev. Phys. Chem.* 31, 401

Weber, P. C., Salemme, F. R., Lin, S. H., Konishi, Y. and Scheraga, H. A. (1985) *J. Mol. Biol.* 181, 453

Wlodawer, A. (1985) in *Biological Macromolecules and Assemblies: Volume 2, Nucleic Acids and Interactive Proteins* (Jurnak, F. A. and McPherson, A., eds), p. 394. New York: Wiley

Wood, W. W. and Erpenbeck, J. J. (1976) *Ann. Rev. Phys. Chem.* 27, 319

Yu, H., Karplus, M. and Hendrickson, W. A. (1985) *Acta Cryst.* B41, 191

Zucker, U. H. and Schulz, H. (1982) *Acta Cryst.* A38, 563

8

Orbital presentation of chemical reactions

KENICHI FUKUI

Kyoto Institute of Technology and Institute for Fundamental Chemistry, Kyoto

With a view to obtaining a picture for an idealized path of chemical reactions the concept of *intrinsic reaction coordinate* (*IRC*) is introduced within the space of the multidimensional potential energy function of the reacting system. The IRC uniquely determines, in a classical sense, the mode of deformation of chemically reacting molecules. On every point along the IRC an orbital analysis is possible with respect to each of the deformed molecules, the geometry of which is frozen to that in the reacting composite system. The analysis is carried out by constructing orbital pairs by respective unitary transformations of canonical molecular orbitals of the deformed molecules, so as to diagonalize the interaction matrix. Of these orbital pairs one or a few, localized in the reacting domain, play the dominant role in offering a distinct view of bond-forming processes along the IRC. In the process of these orbital analyses, the importance of electron delocalization in the chemical interaction is recognized and stressed.

8.1. Introduction

Complicated chemical phenomena, for example organic chemical reactions, are probably one of the most distant 'lands' from the Schrödinger equation in the vast world of its appications. In fact, the chemical reaction is the field which was forsaken by theoretical physicists. In the early stage of the development of applied quantum mechanics, many of the best theorists turned their back upon complex chemical reactions principally on account of the laboriousness involved in computations. Since then, the quantum-mechanical interpretation of chemical reactivity has, rather than otherwise, been the object of theoretical chemists.

8.2. Path of chemical reactions

The first step in the theoretical study of chemical reactions is to set up the concept of the reaction path. For this purpose the model for the 'potential energy surface' of the reacting system is necessary. Modern *ab initio* quantum-chemical methods are capable, on the basis of the Born–Oppenheimer approximation, of calculating the potential energy function, V, in terms of the atomic positions (in cartesian coordinates) X_α, Y_α, Z_α ($\alpha = 1, 2, \ldots, N$), in which N is the number of atoms in the reacting system. The equilibrium points satisfy the simultaneous equations

$$\frac{\partial V}{\partial X_\alpha} = \frac{\partial V}{\partial Y_\alpha} = \frac{\partial V}{\partial Z_\alpha} = 0 \tag{2.1}$$

which include the unique point 'transition state' as well as the reactant and product points. The concept of transition state was early established by Eyring (Eyring, Walter and Kimball, 1944; Glasstone, Laidler and Eyring, 1941; Laidler and Tweedale, 1971) and Polyani (1937). The idealized reaction path should pass through this point. Different approaches have been developed by a number of pioneers to define the reaction path (Basilevsky and Ryabov, 1980; Fernandez and Sinanoglu, 1984; Hofacker, 1969; Jug, 1980; Komornicki *et al.*, 1977; Light, 1971; McIver and Komornicki, 1972; McNutt and Wyatt, 1979; Marcus, 1966*a,b*, 1968; Mezey, 1980, 1981*a,b*, 1982, 1983*a–d*; Millers, 1974; Morokuma and Karplus, 1971; Muller, 1980; Muller and Brown, 1979; Murrell, 1977; Nalewajski, 1978; Porter, 1974; Russegger, 1977, 1978, 1979; Russegger and Brickmann, 1977; Sana, Reckinger and Leroy, 1981; Schaefer, 1979; Walker and Light, 1980; Yarkony, Hunt and Schaefer, 1973). Here, however, the discussion is focussed on the *intrinsic reaction coordinate* or IRC approach (Fukui, 1970*a*, 1974, 1979, 1981*a,b*; Fukui, Kato and Fujimoto, 1975; Fukui, Tachibana and Yamashita, 1981; Kato and Fukui, 1976; Tachibana and Fukui, 1978).

The IRC is a concept which is based on the idea of classical motion of nuclei from the transition state with an infinitesimal velocity. The incremental change of nuclei, dX_α, dY_α, and dZ_α ($\alpha = 1, 2, \ldots, N$) by this 'intrinsic motion' should then satisfy (Fukui, 1974) the simultaneous equations

$$\ldots = \frac{M_\alpha \, dX_\alpha}{\partial V / \partial X_\alpha} = \frac{M_\alpha \, dY_\alpha}{\partial V / \partial Y_\alpha} = \frac{M_\alpha \, dZ_\alpha}{\partial V / \partial Z_\alpha} = \ldots \tag{2.2}$$

where M_α is the mass of nucleus α. Adopting the 'mass-weighted' cartesian coordinates, x_i ($i = 1, 2, \ldots, 3N$) defined by

$$M_\alpha^{1/2} X_\alpha = x_{3\alpha-2}, \qquad M_\alpha^{1/2} Y_\alpha = x_{3\alpha-1}, \qquad M_\alpha^{1/2} Z_\alpha = x_{3\alpha} \tag{2.3}$$

(2.2) is reduced to

$$\frac{\mathrm{d}x_i}{\mathrm{d}s}\frac{\mathrm{d}V}{\mathrm{d}s}=\frac{\partial V}{\partial x_i} \qquad (i=1,2,\dots,3N) \qquad (2.4)$$

with the use of a new parameter s, $\mathrm{d}s$ being the infinitesimal distance of the mass-weighted cartesian configuration space which satisfies (Fukui, 1981b; Kato and Fukui, 1976)

$$\mathrm{d}s^2 = \sum_{i=1}^{3N} \mathrm{d}x_i^2 = 2T\,\mathrm{d}t^2 \qquad (2.5)$$

where T is the kinetic energy and t is the time.
The variational equation

$$\delta\left(\frac{\mathrm{d}V}{\mathrm{d}s}\right)=0 \qquad (2.6)$$

gives at once (2.4), that is the direction of IRC. This implies that the IRC is the steepest descent path in the space of (2.5).

Equation (2.2) or (2.4) has an infinite number of solution curves, $x_i = x_i(s)$ $(i=1,2,\dots,3N)$, of which the IRC is the one passing through the transition state and reaching an equilibrium point. In this manner, the IRC represents the vibrationless–rotationless motion path of the reacting system, and can be obtained numerically. An actual IRC plotting was made concerning several simple reactions (Fukui, Kato and Fujimoto, 1975; Ishida, Morokuma and Komornicki, 1977; Joshi and Morokuma, 1977; Kato and Fukui, 1976; Kato and Morokuma, 1980; Morokuma, Kato and Hirao, 1980).

Once the reaction path is made calculable, the next thing to do is the theoretical selection of the most favourable path among several possible ones with respect to a given reacting system. This constitutes essentially the theory of chemical reactivity, which has for long been the object of interest in the field of organic chemistry.

8.3. Orbitals in chemical reactions

The general discussion of the theory of chemical interaction is not reproduced here, since there are number of papers and books available (Bader, 1962; Devaquet and Salem, 1969; Fujimoto, Inagaki and Fukui, 1976; Fukui, 1964, 1970b, 1971; Fukui, Kato and Yonezawa, 1961; Fukui, Yonezawa and Nagata, 1954; Fukui, Yonezawa and Shingu, 1952; Fukui, Yonezawa, Nagata and Shingu, 1954; Herndon, 1972; Houk, 1975; Hudson, 1973; Klopman, 1968, 1974; Mulliken, 1952, 1956; Pearson, 1976; Picket, Muller and Mulliken, 1953; Salem, 1968, 1969; Woodward and Hoffmann, 1965, 1969a,b, 1974). The importance of electron delocalization between reactants as a major driving force of chemical reactions was specially mentioned in these references.

The molecular orbitals (MOs) which are specified as often playing a · dominant role in the delocalization process were called *frontier orbitals* (FOs) (Fukui, 1970*b*; Fukui, Yonezawa and Shingu, 1952). The FOs were originally defined as the highest occupied (HO) MO for the *electrophilic* reaction and the lowest unoccupied (LU) MO for the *nucleophilic* reaction. Various sorts of chemical reactions of various chemical compounds were interpreted by aid of HOMO–LUMO interaction scheme (Fukui, 1970*b*).

It turned out that in most usual cases the FO pattern determines the most favourable IRC path. In these circumstances, the orbital presentation is also attempted by laying stress on the frontier orbital approach.

8.4. Interaction frontier orbitals

The electron delocalization yields accumulation of the electron density in the intermolecular region between reaction sites. A succinct description of the delocalization mode, which involves the contribution of all the MOs, is possible by a pair of transformations of the canonical MOs (Hartree–Fock–Roothaan MOs) of the fragments A and B in a composite reacting system (Fukui, Koga and Fujimoto, 1981). The geometries of the fragments A and B are frozen to be the same as those in the reacting system A–B. The geometry of the whole reacting system can be obtained from the IRC approach.

The 'delocalization energy matrix' $\mathbf{D}$ is defined by the interaction energy, D_{il}, associated with the electron transfer between the ith occupied MO of the donor fragment, say A, and the lth unoccupied MO of the acceptor fragment, say B. The product $\mathbf{D}^{+}\mathbf{D}$ is hermitian and is diagonalized by a unitary transformation $\mathbf{U}$, so that

$$\mathbf{D}^{+}\mathbf{D}\mathbf{U} = \mathbf{U}\mathbf{\Lambda} \qquad (4.1)$$

in which the non-negative eigenvalue matrix is $\mathbf{\Lambda}$.

By the use of this $\mathbf{U}$, one obtains two sets of transformations among the occupied MOs of the donor and among the unoccupied MOs of the acceptor which convert the canonical MOs into a set of paired hybrids of the fragments. The procedure is similar to that of the Amos–Hall corresponding orbital (Amos and Hall, 1961). However, unlike the Amos–Hall case, one has to repeat the transformation to reach the true pairing of orbitals. Each set of the new orbitals constitutes an orthonormal set, retaining the electronic structures of the fragments unchanged. It is shown that FOs make dominant contributions to the interaction molecular hybrids. In the case of proton interaction, for instance, one obtains a single donor hybrid exclusively taking part in the protonation. In this way, one can condense the entire consequence of delocalization interaction into a pair or, at most, a few pairs of molecular hybrids of fragments. These pairs are remarkably localized in the region of interaction.

The most dominant pair of hybrids may be called *interaction frontier orbitals* (IFOs), since these hybrids are importantly localized in the *frontiers* of mutual interaction and, in a great number of cases, also energetically correspond to the hybridized HOMO or LUMO of interacting molecules. If desired, the IFOs are, of course, similarly obtained with respect to the unoccupied orbitals of the donor and the occupied orbitals of the acceptor. Therefore, the donor–acceptor argument is essentially unnecessary.

8.5. Interactive hybrid orbitals

A modification can be made by replacing the delocalization energy matrix with the 'intermolecular overlap population matrix', $\mathbf{P}$ (Fujimoto *et al.*, 1981*a*). By diagonalizing the product $\mathbf{P}^+\mathbf{P}$, one obtains *interactive hybrid orbitals* (IHOs). The diagonalization is achieved by two sets of repeated unitary transformations among all the MOs of reactant A and among all the MOs of reactant B. The IHOs do not have to be either occupied or unoccupied. They are something in-between, depending on the type of interaction. Several examples for IHOs are shown in a few connected papers (Fujimoto and Koga, 1982; Fujimoto, Koga and Fukui, 1981; Fujimoto *et al.*, 1981*b*, 1982, 1983, 1984). The results of calculation are, however, not so much different from those of the preceding method. A merit of this modified method is to make obtainable the contributions from all the orbital effects, not limited to that of delocalization.

8.6. Further generalizations

An extensive generalization on this line is possible. The MOs, Φ_m, of the composite system A–B is expanded in terms of the MOs of fragments A and B, ϕ_i and ψ_k, respectively, as

$$\Phi_m = \sum_i^A C_{m,i}\phi_i + \sum_k^B C_{m,k}\psi_k \qquad (6.1)$$

The *intermolecular interaction matrix* $\mathbf{P}$ is defined in regard to any appropriate one-electron operator $\mathbf{p}(1)$

$$\mathbf{P} = (P_{i,k}) \qquad (6.2)$$

For the case of a closed-shell electronic structure represented by a single-determinant wavefunction

$$P_{i,k} = 4 \sum_{m}^{\text{occ}} c_{m,i} c_{m,k} \int \phi_i(1) \mathbf{p}(1) \psi_k(1) \, dv_1 \qquad (6.3)$$

diagonalization of the product $\mathbf{P}^+\mathbf{P}$ gives non-negative eigenvalues. One of the new pairs of orbitals, say ϕ'_f and ψ'_f, which correspond to the largest eigenvalue, should have the greatest amplitude in the region of mutual interaction in A and that in B, respectively (Fujimoto, Koga and Fukui, 1981; Fujimoto *et al.*, 1984).

When the matrix $\mathbf{P}$ spans only the occupied subset of donor orbitals and the unoccupied subset of acceptor orbitals, the orbital transformations yield IFO-like hybrids. When the matrix $\mathbf{P}$ spans the set of all MOs of reactant A and the set of all MOs of reactant B, the orbital transformations result in IHO-like orbitals. An example of unrestricted Hartree–Fock SCF MO calculation is also given (Fujimoto, Koga and Fukui, 1981; Fujimoto *et al.*, 1984).

In combination with the configuration analysis technique (Baba, Suzuki and Takemura, 1969; Fujimoto *et al.*, 1974, Fukui and Fujimoto, 1968; Fukui, Kato and Fujimoto, 1975) with respect to the wavefunction of interacting systems, the presentation of the effect of electron delocalization is much simplified (Fujimoto, Koga and Hataue, 1984; Fujimoto *et al.*, 1985*a*). The ground-state wavefunction of the composite reacting system A–B, Ψ, is presented by a linear combination of various electron configurations, Ψ_p, of fragments A and B (Fujimoto *et al.*, 1974; Fukui, Kato and Fujimoto, 1975). These electron configurations cover the original one, Ψ_0, electron-transferred ones, Ψ_{i-l} and Ψ_{k-j}, in which i and k signify the occupied MOs of A and B, respectively, and j and l indicate unoccupied MOs of A and B, respectively, and locally excited ones, Ψ_{i-j} and Ψ_{k-l}. Besides, many other less important configurations are possibly able to mix in.

The MOs of A–B are denoted by Φ_f, where $f = 1, 2, \ldots, m+n$, m and n being the number of occupied MOs of A and of B, respectively. The MO Φ_f can be expanded in terms of the fragment MOs ϕ of A and ψ of B, as

$$\Phi_f = \sum_{i=1}^{m} c_{i,f} \phi_i + \sum_{j=1}^{M-m} c_{m+j,f} \phi_{m+j} + \sum_{k=1}^{n} d_{k,f} \psi_k + \sum_{l=1}^{N-n} d_{n+l,f} \psi_{n+l} \qquad (6.4)$$

$(f = 1, 2, \ldots, m+n)$ or simply written as

$$(\boldsymbol{\Phi}) = (\phi^{\text{occ}} \phi^{\text{uno}} \psi^{\text{occ}} \psi^{\text{uno}}) \begin{pmatrix} \mathbf{c}^{\text{occ}A} \\ \mathbf{c}^{\text{uno}A} \\ \mathbf{d}^{\text{occ}B} \\ \mathbf{d}^{\text{uno}B} \end{pmatrix}$$

Kenichi Fukui

where M and N represent the number of basis functions on A and on B, respectively, and accordingly $M - m$ and $N - n$ are the number of unoccupied MOs of A and of B.

In this analysis a few assumptions are made for simplicity: (1) all the atomic integrals bridging A and B are disregarded in MO calculation; (2) the reacting system A–B has a closed-shell electronic structure with the wavefunction represented by a single Slater determinant; and also (3) $m > N - n$. Removal of conditions (2) and (3) for generalization is easy.

Then, the wavefunction of A–B, denoted by Ψ, is represented by

$$\Psi = |\Phi_1(1)\bar{\Phi}_1(2) \cdots \Phi_{m+n}(2m+2n-1)\bar{\Phi}_{m+n}(2m+2n)|$$

$$< \sum_p C_p \Psi_p \tag{6.5}$$

in which Ψ_p is the term corresponding to the electron configuration p of the reacting system A–B. The function Ψ_p is easily obtained in terms of the fragment MOs (Fujimoto *et al.*, 1974).

A technique similar to the previous pairwise unitary transformations with regard to the occupied subset of A and the unoccupied subset of B can be applied, so that one can assure the relation

$$\begin{aligned} C_{i,n+1} &\neq 0 \qquad \text{when } i = 1 \\ &= 0 \qquad \text{when } i \neq 1 \end{aligned} \tag{6.6}$$

after convergence is attained. This can be carried out by diagonalizing the hermitian matrix $(\mathbf{V}^{occA})^+ \mathbf{V}^{occA}$ in which $\mathbf{V}^{occA}$ satisfies

$$\mathbf{d}^{unoB} = (\mathbf{V}^{occA} \, \mathbf{V}^{occB}) \begin{pmatrix} \mathbf{c}^{occA} \\ \mathbf{d}^{occB} \end{pmatrix} \tag{6.7}$$

Relation (6.6) implies that the electron delocalization takes place only between the paired orbitals after transformation, ϕ'_1 (doubly occupied) and ψ'_{n+1} (unoccupied). The total amount of delocalized electron density transferred from A to B is kept unchanged by the paired unitary transformations. Therefore, this procedure gives the most succinct and precise orbital presentation of the electron delocalization between chemically interacting species. In the case of proton addition to acrolein, a calculation shows that the HO and the next HO MOs of acrolein are the major constituents of the IFO.

8.7. Complementary remarks

A simplified version of the IFO approach for single-site reactions, which affords a hybridized MO which is most strongly localized at the reaction site, arises when a unitary transformation is applied to the occupied

MOs of a reactant molecule for an electrophilic reaction, and to the unoccupied MOs for a nucleophilic reaction (Fujimoto, Mizutani and Iwase, 1986). Obviously, this orbital, named *localized reactive orbitals* (LROs), gives the maximum overlap with the relevant reagent orbitals at the reaction site. A typical example of single-site reactions suitable for the present approach is well-examined aromatic substitutions. An excellent parallelism with experimental results is found between the LRO energy and the chemical reactivity. The smaller the LRO energy, the greater is the reactivity. An extension to the case of multi-centred reactions is easy.

It should be noted that LROs are the most *de*localizable orbitals, and that the beautiful correlation of the LRO result with experience implies nothing but the significance of the electron *de*localization process in chemical reactions.

An interesting relation between the chemical potential of an assembly of electrons in the molecular state and the FO density function was reported by Parr and Yang (1984). This important result suggests a possibility that the FO theory is connected with the thermodynamic property of the assembly of molecular electrons through their delocalizing ('escaping') tendency.

References

Amos, A. T. and Hall, G. G. (1961) *Proc. Roy. Soc. London Ser. A* 263, 483

Baba, H., Suzuki, S. and Takemura, T. (1969) *J. Chem. Phys.* 50, 2078

Bader, R. F. W. (1962) *Can. J. Chem.* 40, 1164

Basilevsky, M. V. and Ryabov, V. M. (1980) *Chem. Phys.* 50, 231

Devaquet, A. and Salem, L. (1969) *J. Am. Chem. Soc.* 91, 3743

Eyring, H., Walter, J. and Kimball, G. E. (1944) *Quantum Chemistry.* John Wiley, New York

Fernandez, A. and Sinanoglu, O. (1984) *Theor. Chim. Acta (Berl.)* 65, 179

Fujimoto, H. and Fukui, K. (1974) in *Chemical Reactivity and Reaction Paths* (Klopman, G., ed.), p. 23. Wiley-Interscience, New York

Fujimoto, H. and Koga, N. (1982) *Tetrahedron Lett.* 24, 4357

Fujimoto, H. and Yamasaki, T. (1986) *J. Am. Chem. Soc.* 108, 578

Fujimoto, H., Inagaki, S. and Fukui, K. (1976) *J. Am. Chem. Soc.* 98, 2670

Fujimoto, H., Koga, N. and Hataue, I. (1984) *J. Phys. Chem.* 88, 3539

Fujimoto, H., Koga, N. and Fukui, K. (1981) *J. Am. Chem. Soc.* 103, 7452

Fujimoto, H., Mizutani, Y. and Iwase, K. (1986) *J. Phys. Chem.* (in press)

Fujimoto, H., Endo, M., Hataue, I. and Koga, N. (1982) *Tetrahedron Lett.* 23, 5559

Fujimoto, H., Hataue, I., Koga, N. and Yamasaki, T. (1984) *Tetrahedron Lett.* 25, 5339

Fujimoto, H., Kato, S., Yamabe, S. and Fukui, K. (1974) *J. Chem. Phys.* 60, 572

Fujimoto, H., Koga, N., Endo, M. and Fukui, K. (1981a) *Tetrahedron Lett.* 22, 1263

Fujimoto, H., Koga, N., Endo, M. and Fukui, K. (1981b) *Tetrahedron Lett.* 22, 3427

Fujimoto, H., Yamasaki, T., Hataue, I. and Koga, N. (1985a) *J. Phys. Chem.* 89, 779

Fujimoto, J., Yamasaki, T., Mizutani, H. and Koga, N. (1985b) *J. Am. Chem.* 88, 3539

Fujimoto, H., Koga, N., Endo, M., Hataue, I. and Fukui, K. (1983) *Israel J. Chem.* 23, 49

Fukui, K. (1964) in *Molecular Orbitals in Chemistry* (Löwdin, P-O. and Pullman, B., eds), p. 513. Academic Press, New York

Fukui, K. (1970a) *J. Phys. Chem.* 74, 4161

Fukui, K. (1970b) *Theory of Orientation and Stereoselection.* Springer, Berlin

Fukui, K. (1971) *Acc. Chem. Res.* 4, 57

Fukui, K. (1974) in *The World of Quantum Chemistry.* Proceedings of the First International Congress of Quantum Chemistry, Menton, 1973 (Daubel, R. and Pullman, B., eds), p. 113. D. Reidel, Dordrecht

Fukui, K. (1979) *Recl. Trav. Chim. Pays-Bas* 98, 75

Fukui, K. (1981a) *Intern. J. Quant. Chem. Quant. Chem. Symp.* 15, 633

Fukui, K. (1981b) *Acc. Chem. Res.* 14, 363

Fukui, K. and Fujimoto, H. (1968) *Bull. Chem. Soc. Japan* 41, 1989

Fukui, K., Kato, S. and Fujimoto, H. (1975) *J. Am. Chem. Soc.* 97, 1

Fukui, K., Kato, H. and Yonezawa, T. (1961) *Bull. Chem. Soc. Japan* 34, 1112

Fukui, K., Koga, N. and Fujimoto, H. (1981) *J. Am. Chem. Soc.* 103, 196

Fukui, K., Tachibana, A. and Yamashita, K. (1981) *Intern. J. Quant. Chem. Quant. Chem. Symp.* 15, 621

Fukui, K., Yonezawa, T. and Nagata, C. (1954) *Bull. Chem. Soc. Japan* 27, 423

Fukui, K., Yonezawa, T. and Shingu, H. (1952) *J. Chem. Phys.* 20, 722

Fukui, K., Yonezawa, T., Nagata, C. and Shingu, H. (1954) *J. Chem. Phys.*, 22, 1433

Glasstone, S., Laidler, K. and Eyring, H. (1941) *The Theory of Rate Processes.* McGraw-Hill, New York

Herndon, W. C. (1972) *Chem. Rev.* 72, 157

Hofacker, L. (1969) *Intern. J. Quant. Chem.* 35, 33

Houk, K. N. (1975) *Acc. Chem. Res.* 8, 361

Hudson, R. F. (1973) *Angew. Chem. Int. Ed. Eng.* 12, 36

Ishida, K., Morokuma, K. and Komornicki, A. (1977) *J. Chem. Phys.* 66, 2153

Joshi, B. D. and Morokuma, K. (1977) *J. Chem. Phys.* 67, 4880

Jug, K. (1980) *Chem. Phys.* 54, 263

Kato, S. and Fukui, K. (1976) *J. Am. Chem. Soc.* 98, 6395

Kato, S. and Morokuma, K. (1980) *J. Chem. Phys.* 73, 3900

Klopman, G. (1968) *J. Am. Chem. Soc.* 90, 223

Klopman, G. (ed.) (1974) *Chemical Reactivity and Reaction Paths.* John Wiley, New York

Komornicki, A., Ishida, K., Morokuma, K., Ditchfield, R. and Conrad, M. (1977) *Chem. Phys. Lett.* 45, 595

Laidler, J. and Tweedale, A. (1971) *Adv. Chem. Phys.* 21, 113
Light, J. C. (1971) *Adv. Chem. Phys.* 19, 1
McIver, J. W. Jr. and Komornicki, J. (1972) *J. Am. Chem. Soc.* 94, 2625
McNutt, J. F. and Wyatt, R. E. (1979) *J. Chem. Phys.* 70, 5307
Marcus, R.A. (1966a) *J. Chem. Phys.* 45, 4493
Marcus, R. A. (1966b) *J. Chem. Phys.* 45, 2138, 2630
Marcus, R. A. (1968) *J. Chem. Phys.* 49, 2610
Mezey, P. G. (1980) *Theor. Chim. Acta (Berl.)* 54, 95
Mezey, P. G. (1981a) *Theor. Chim. Acta (Berl.)* 58, 309
Mezey, P. G. (1981b) *Intern. J. Quant. Chem. Quant. Biol. Symp.* 8, 185
Mezey, P. G. (1982) *Theor. Chim. Acta (Berl.)* 60, 409
Mezey, P. G. (1983a) *Intern. J. Quant. Chem. Quant. Chem. Symp.* 17, 137, 453
Mezey, P. G. (1983b) *J. Chem. Phys.* 78, 6182
Mezey, P. G. (1983c) *Can. J. Chem.* 61, 956
Mezey, P. G. (1983d) *J. Mol. Struct.* 103, 81
Millers, W. H. (1974) *J. Chem. Phys.* 61, 1823
Morokuma, K. and Karplus, M. (1971) *J. Chem. Phys.* 55, 63
Morokuma, K., Kato, S. and Hirao, K. (1980) *J. Chem. Phys.* 72, 6800
Müller, K. (1980) *Angew. Chem. Int. Ed. Eng.* 19, 1
Müller, K. and Brown, L. D. (1979) *Theor. Chim. Acta (Berl.)* 53, 75
Mulliken, R. S. (1952) *J. Am. Chem. Soc.* 74, 811
Mulliken, R. S. (1956) *Rec. Trav. Chim.* 75, 845
Murrell, J. N. (1977) *Struct. Bonding (Berl.)* 32, 93
Nalewajski, R. F. (1978) *Intern. J. Quant. Chem.* 12, 87
Parr, R. G. and Yang, W. (1984) *J. Am. Chem. Soc.* 106, 4049
Pearson, R. G. (1976) *Symmetry Rules for Chemical Reactions.* John Wiley, New York
Picket, L. W., Muller, N. and Mulliken, R. S. (1953) *J. Chem. Phys.* 21, 1400
Polyani, M. (1937) *J. Chem. Soc.* p. 629
Porter, R. N. (1974) *Ann. Rev. Phys. Chem.* 25, 317
Russegger, P. (1977) *Chem. Phys.* 22, 41
Russegger, P. (1978) *Chem. Phys.* 34, 329
Russegger, P. (1979) *Chem. Phys.* 41, 299
Russegger, P. and Brickmann, J. (1977) *J. Chem. Phys.* 66, 1
Sana, M., Reckinger, G. and Leroy, G. (1981) *Chem. Phys.* 58, 145
Salem, L. (1968) *J. Am. Chem. Soc.* 90, 543, 553
Salem, L. (1996) *Chem. Brit.* 5, 449
Schaefer, H. F. (1979) in *Atom–Molecule Collision Theory: A Guide for the Experimentalist* (Bernstein, R. B., ed.), p. 45. Plenum, New York
Tachibana, A. and Fukui, K. (1978) *Theor. Chim. Acta (Berl.)* 49, 321
Walker, R. B. and Light, J. C. (1980) *Ann. Rev. Phys. Chem.* 31, 401
Woodward, R. B. and Hoffman, R. (1965) *J. Am. Chem. Soc.* 98, 2670
Woodward, R. B. and Hoffmann, R. (1969a) *Angew. Chem.* 81, 797
Woodward, R. B. and Hoffmann, R. (1969b) *The Conservation of Orbital Symmetry.* Academic Press, New York
Woodward, R. B. and Hoffmann, R. (1974) see papers in *Orbital Symmetry Papers* (Simmons, H. E. and Bunnett, J. F., eds). ACS, Washington, DC
Yarkony, D. R., Hunt, W. J. and Schaefer, H. F. (1973) *Mol. Phys.* 26, 941

9

Quantum chemistry

A.D. BUCKINGHAM

University Chemical Laboratory, Cambridge

Quantum mechanics is crucial to an understanding of chemistry. As Linus Pauling (1985) put it: 'Chemistry is a quantum phenomenon, or, rather, a great collection of quantum phenomena'. The name of Schrödinger is as familiar to present-day chemistry undergraduates as is that of Faraday, Kekulé or Mendeléev. There exists a thriving branch of the subject known as quantum chemistry; it is concerned with approximate solutions of the Schrödinger equation $\mathscr{H}\Psi_n = E_n\Psi_n$ and it provides descriptions of atoms, molecules and clusters that are of interest to chemists. But there are still a number of scientists – and great chemists among them – who are of the opinion that they do not need a knowledge of quantum mechanics. They believe that the results of quantum-chemical computations represent a simplified description of reality that may miss the essential truth. It has sometimes been suggested that results obtained by computation are lacking in elegance and are less important than those deduced by pure reasoning (Coulson, 1960). Hirschfelder (1983) wrote that 'scientists in the 1980s get so immersed in a maze of computational detail that they lose sight of the simple, elegant theories.'

It is true that computations may have a short life; we have been going through a phase where *ab initio* calculations on small molecules may be improved annually!

But through computation, quantum chemists have created (Davidson, 1984) 'a quantitative model of the chemical bond which is beautiful to those who understand it and which is likely to be permanent. Further, this development would never have been possible by reasoning alone.'

Chemists are concerned with the forces that hold atoms together – the chemical bond. The early theoretical era employed concepts such as the shared electron pair, the octet rule, electronegativity and ionic bonds. This approach was expounded most beautifully in *The Nature of the Chemical Bond* by Pauling (1940), in which the old chemical concepts were interpreted in terms of wavefunctions. Predictions of chemical properties were made on the basis of interpolation and extrapolation from known facts. Pauling's predictions of bond-lengths, heats of reaction, etc. had a profound effect on twentieth century chemistry; their reliability testifies to Pauling's genius, but such an approach has obvious limitations.

The molecular-orbital model of the chemical bond arose through the word of Hund (1928), Mulliken (1932, 1978), Lennard-Jones (1929, 1949), Walsh (1953) and others. As a qualitative tool this model also led to predictions of bond lengths and angles and of electronic spectra of simple molecules, as well as to the celebrated Woodward–Hoffmann rules describing chemical reactivity (Woodward and Hoffmann, 1970). The molecular-orbital model has been refined and developed from crude semi-empirical theories such as Hückel's approach to molecules containing double bonds (Hückel, 1931), to the more elaborate approaches known as extended Hückel theory (Hoffmann, 1968), CNDO developed by Pople (Pople and Beveridge, 1970), and MNDO due to Dewar (Dewar and Thiel, 1977).

Until about 1950, it seemed that we would have to be content with semi-empirical theories, except for the simplest molecules, such as H_2 for which James and Coolidge (1933, 1935) used a 13-term variational wavefunction for the ground state, containing the interelectronic distance r_{12} directly, as in the pioneering work of Hylleraas (1928, 1964) on the helium atom. The computational bottleneck began to be broken with the development of electronic digital computers, and methods for evaluating the difficult electron-repulsion integrals were developed by Kotani, Coulson, Löwdin, Slater, Roothaan and particularly by Boys (1950). The mathematical formalism for rigorous solution for the self-consistent-field (SCF) equations in terms of a finite basis of one-electron functions was established by Roothaan (1951) and Hall (1951).

An alternative approach to the electronic structure of molecules emerged from the work of Heitler and London (1927) on the H_2 molecule; McCrea (1985) writes that

'this was the most important problem I considered in those days, but I got nowhere. Then one day in 1927, I was able to tell Fowler that a paper by Walter Heitler and Fritz London apparently solved the problem in terms of a new concept: a quantum-mechanical exchange force. He grasped the idea at once, and bade me expound it at the next colloquium – which is how quantum chemistry came to Britain.'

This approach was called the valence bond (VB) theory and it played a major role in the development of our ideas about the chemical bond (Pauling, 1940). The VB theory proved more difficult to reduce to a simple computer algorithm than was the case for the molecular orbital (MO) theory, but there aré indications that the approach is now becoming more competitive (Cooper, Gerrat and Raimondi, 1986).

The development of the present powerful computer programs for the computation of the electronic structure and properties of molecules has been described by some leading practitioners, for example, Pople (Hehre *et al.*, 1986), Schaefer (1984), Davidson (1984) and Handy (1984). We have reached the stage where *ab initio* computations provide a very important source of reliable and detailed information about molecules in their ground and excited states. The programs are widely available, and experimentalists are using them to complement their measurements. A few examples will be mentioned to illustrate the accuracy that can be achieved in quantum-chemical computations.

The dissociation energy D_0 of the hydrogen molecule was computed by Kołos and Wolniewicz (1968) to be $36117.4 \, \mathrm{cm}^{-1}$, which was larger than the then accepted experimental value of $36113.6 \pm 0.3 \, \mathrm{cm}^{-1}$ (Herzberg and Morfils, 1960). The discrepancy was resolved by a new analysis of the spectroscopic data to yield the experimental value $D_0(\mathrm{H}_2) = 36117 \pm 1.0 \, \mathrm{cm}^{-1}$ (Herzberg, 1970). It is of interest to note that the non-relativistic Born–Oppenheimer 'clamped nucleus' approximation gives a very accurate description of the bonding in H_2 (it would be even better in D_2), for the non-adiabatic effects due to coupling of the electronic and nuclear motions contribute only about $0.5 \, \mathrm{cm}^{-1}$ to $D_0(\mathrm{H}_2)$ and relativistic and radiative effects $-0.7 \, \mathrm{cm}^{-1}$ (Kołos and Wolniewicz, 1968).

The polarizability α of the helium atom was calculated (Buckingham and Hibbard, 1968) to be 1.38319 atomic units by using very accurate variational wavefunctions, and rather tight upper and lower bounds were imposed upon this result (Glover and Weinhold, 1976). This was far more accurate than experiment at the time, but improved techniques of dielectric constant gas thermometry led to a value of α which seemed to be significantly larger than the computed value (Gugan and Michel, 1980). However, the discrepancy was due to a conversion of the computation to practical units, for the finite mass of the He nucleus should be allowed for, leading to a computed $\alpha(\mathrm{He}) = 0.205051 \times 10^{-24} \, \mathrm{cm}^3$ (Weinhold, 1982) to be compared with an experimental value of $0.205057 \pm 0.000010 \times 10^{-24} \, \mathrm{cm}^3$ (Gugan and Michel, 1980). The computation on the helium atom in an electric field was extended to higher order in the field E, giving the static hyperpolarizability of helium (i.e. the energy term in the fourth power of E) to four significant figures (Buckingham and Hibbard, 1968).

Molecular interaction energies can be calculated by quantum-chemical techniques (Pople, 1982; van Lenthe, van Dam and van Duijneveldt, 1984), but there can be difficulties due to 'basis set superposition errors' resulting from the effective expansion of the size of the basis set when the two molecules interact. Boys and Bernardi (1970) introduced the 'counterpoise' technique to counteract this problem. It certainly helps (Fowler and Buckingham, 1983), but there is still a controversy relating to the reliability of this approach (Collins and Gallup, 1986; Gutowski *et al.*, 1986; Schwenke and Truhlar, 1985).

Quantum-chemical computations can be instructive, and possibly more reliable than experiment, in the case of a property like the dipole moment of GeH. Intensity measurements in laser-magnetic resonance spectroscopy (Brown, Evenson and Sears, 1985) indicated that the dipole moment of GeH in the ground $^2\Pi$ state is 1.24 ± 0.1 debye. This is to be compared with the dipole moments of CH (1.40 debye in the sense C^-H^+) and SiH (0.09 debye, Si^-H^+) (Meyer and Rosmus, 1975), so it seemed likely that the sense of the dipole in GeH is Ge^+H^-. Large-scale SCF calculations with configuration interaction (Pettersson and Langhoff, 1986) and with the coupled electron pair approximation (Meyer, 1973; Werner and Buckingham, 1986) indicate that the dipole moment is only ~ 0.1 debye (Ge^+H^-), so it will be interesting to learn what value will be provided by accurate Stark splittings.

It is now routine to calculate potential energy surfaces for molecules of interest to organic chemists. Such surfaces not only give information about the bond lengths and angles of the equilibrium structure but also provide the heights of potential barriers to chemical rearrangements (Hehre *et al.*, 1986). The direct evaluation of analytic derivatives of the energy, and of properties such as the dipole moment, initiated by Gerratt and Mills (1968), Pulay (1969, 1970, 1971), Pople (Pople *et al.*, 1979), Schaefer and Handy (Lee *et al.*, 1986), has proved to be an efficient procedure for determining equilibrium structures, harmonic frequencies, vibrational intensities, and transition states (which are not accessible to experimental observation); with large basis sets and with electron correlation taken to the second order of perturbation theory (Møller and Plesset, 1934) it is possible to predict single-bond lengths to about ± 0.002 Å, multiple-bond lengths to ± 0.01 Å, angles to $\pm 0.2°$ and harmonic frequencies to 1.5% for molecules of the size of H_2CO (Simandiras, Handy and Amos, 1986).

It is clear from the above that modern quantum chemistry has much to offer. But major problems remain. In the case of medium-sized molecules, basis-set deficiencies can be severe – it would be nice to escape from this limitation, as has been done in obtaining solutions of the Hartree–Fock equations by numerical means in the case of linear molecules (Laaksonen, Pyykkö and Sundholm, 1983; McCullough, 1982); however, the multidimensionality of molecules appears to provide a formidable barrier to progress in this direction.

Another major problem is to be found in the case of transition-metal compounds. Electron correlation seems to be of greater relative importance for these systems, presumably because of the large number of low-lying atomic energy levels. Thus semi-empirical approaches, such as crystal-field theory (Tanabe and Sugano, 1954), the angular-overlap model (Schäffer and Jorgensen, 1965; Woolley, 1987), and the SCFXα method (Slater, 1979) have held sway.

While much remains to be learned, there can be no doubt that Schrödinger's equation, coupled with the Born–Oppenheimer separation of the electronic and nuclear motions (Born and Huang, 1954; Born and Oppenheimer, 1927), provides the theoretical basis for a sound description of chemistry and of molecular biology. The prognosis is good for quantum chemistry as it moves from its youth towards full maturity.

References

Born, M. and Huang, K. (1954) *Dynamical Theory of Crystal Lattices*, p. 166. Clarendon Press, Oxford

Born, M. and Oppenheimer, R. (1927) *Ann. Phys.* 84, 457

Boys, S. F. (1950) *Proc. Roy. Soc. A* 200, 542

Boys, S. F. and Bernardi, F. (1970) *Mol. Phys.* 19, 553

Brown, J. M., Evenson, K. M. and Sears, T. J. (1985) *J. Chem. Phys.* 83, 3275

Buckingham, A. D. and Hibbard, P. G. (1968) *Symposia Faraday Soc.* 2, 41

Collins, J. R. and Gallup, G. A. (1986) *Chem. Phys. Lett.* 129, 329

Cooper, D. L., Gerratt, J. and Raimondi, M. (1986) *Adv. Chem. Phys.* (in press)

Coulson, C. A. (1960) *Rev. Mod. Phys.* 32, 170

Davidson, E. R. (1984) *Faraday Symposia* 19, 7

Dewar, M. J. S. and Thiel, W. (1977) *J. Am. Chem. Soc.* 99, 4899

Fowler, P. W. and Buckingham, A. D. (1983) *Mol. Phys.* 50, 1349

Gerratt, J. and Mills, I. M. (1968) *J. Chem. Phys.* 49, 1719

Glover, R. M. and Weinhold, F. (1976) *J. Chem. Phys.* 65, 4913

Gugan, D. and Michel, G. W. (1980) *Metrologia* 16, 149

Gutowski, M., van Duijneveldt, F. B., Chalasinski, G. and Pula, L. (1986) *Chem. Phys. Lett.* 129, 325

Hall, G. G. (1951) *Proc. Roy. Soc. A* 205, 541

Handy, N. C. (1984) *Faraday Symposia Chem. Soc.* 19, 17

Hehre, W. J., Radom, L., Schleyer, P. v. R. and Pople, J. A. (1986) *Ab initio Molecular Orbital Theory.* Wiley, New York

Heitler, W. and London, F. (1927) *Z. Phys.* 44, 455

Herzberg, G. (1970) *J. Mol. Spect.* 5, 482

Herzberg, G. and Monfils, A. (1960) *J. Mol. Spect.* 5, 482

Hirschfelder, J. O. (1983) *Ann. Rev. Phys. Chem.* 34, 1
Hoffmann, R. (1968) *J. Chem. Phys.* 39, 1397
Huckel, E. (1931) *Z. Phys.* 70, 204
Hund, .F (1928) *Z. Phys.* 51, 759
Hylleraas, E. A. (1928) *Z. Physik* 48, 469
Hylleraas, E. A. (1964) *Adv. Quantum Chem.* 1, 1
James, H. M. and Coolidge, A. S. (1933) *J. Chem. Phys.* 1, 825
James, H. M. and Coolidge, A. S. (1935) *J. Chem. Phys.* 3, 120
Kołos, W. and Wolniewicz, L. (1968) *J. Chem. Phys.* 49, 404
Laaksonen, L., Pyykkö, P. and Sundholm, D. (1983) *Chem. Phys. Lett.* 96, 1
Lee, T. J., Handy, N. C., Rice, J. E., Scheiner, A. C. and Schaefer, H. F., III
    (1986) *J. Chem. Phys.* 85, 3930
Lennard-Jones, J. E. (1929) *Trans. Faraday Soc.* 25, 668
Lennard-Jones, J. E. (1949) *Proc. Roy. Soc. A* 198, 1
McCrea, W. (1985) *New Scientist*, 17 October, p. 58
McCullough, E. A., Jr (1982) *J. Phys. Chem.* 86, 2178
Meyer, W. (1973) *J. Chem. Phys.* 58, 1017
Meyer, W. and Rosmus, P. (1975) *J. Chem. Phys.* 63, 2356
Møller, C. and Plesset, M. S. (1934) *Phys. Rev.* 46, 618
Mulliken, R. S. (1932) *Phys. Rev.* 40, 55
Mulliken, R. S. (1978) *Ann. Rev. Phys. Chem.* 29, 1
Pauling, L. (1940) *The Nature of the Chemical Bond*, 2nd edn. Cornell
    University Press, Ithaca
Pauling, L. (1985) *New Scientist*, 7 November, p. 54
Pettersson, L. G. M. and Langhoff, S. R. (1986) *Chem. Phys. Lett.* 125, 429
Pople, J. A. (1982) *Faraday Discussions Chem. Soc.* 73, 7
Pople, J. A. and Beveridge, D. L. (1970) *Approximate Molecular Orbital*
    Theory. McGraw-Hill
Pople, J. A., Krishnan, R., Schlegel, H. B. and Binkley, J. S. (1979) *Int. J.*
    Quant. Chem. S13, 225
Pulay, P. (1969) *Mol. Phys.* 17, 197
Pulay, P. (1970) *Mol. Phys.* 18, 473
Pulay, P. (1971) *Mol. Phys.* 21, 329
Roothaan, C. C. J. (1951) *Rev. Mod. Phys.* 23, 69
Schaefer, H. F., III (1984) *Quantum Chemistry: The Development of* Ab Initio
    Methods in Molecular Electronic Structure Theory. Clarendon Press, Oxford
Schäffer, C. E. and Jorgensen, C. K. (1965) *Mol. Phys.* 9, 401
Schwenke, D. W. and Truhlar, D. G. (1985) *J. Chem. Phys.* 82, 2418
Simandiras, E. D., Handy, N. C. and Amos, R. D. (1987) *Chem. Phys. Lett.* (in
    press)
Slater, J. C. (1979) *The Calculation of Molecular Orbitals.* Wiley, New York
Tanabe, Y. and Sugano, S. (1954) *J. Phys. Soc. Japan* 9, 753
van Lenthe, J. H., van Dam, T. and van Duijneveldt, F. B. (1984) *Faraday*
    Symposia Chem. Soc. 19, 125
Walsh, A. D. (1953) *J. Chem. Soc.* p. 2260
Weinhold, F. (1982) *J. Phys. Chem.* 86, 1111
Werner, H-J. and Buckingham, A. D. (1986) *Chem. Phys. Lett.* 125, 433

Woodward, R. B. and Hoffmann, R. (1970) *The Conservation of Orbital Symmetry.* Academic Press, New York

Woolley, R. G. (1987) *Int. Rev. Phys. Chem.* (in press)

10

Eamon de Valera, Erwin Schrödinger and the Dublin Institute

SIR WILLIAM McCREA

University of Sussex

10.1. Introduction

Two of the great men of our times are remembered here. Each possessed a wondrous range of endowments. They were men of courageous vision. The fruits of their endeavour and their vision have much significance in diverse ways for ourselves and our successors. Both were obviously of extraordinary independence of mind and given to much solitary work, yet both owed a great deal to their living among contemporaries who were themselves of outstanding distinction and dedication to their ideals.

Eamon de Valera, statesman and leader, visionary, natural scholar and devotee of mathematics, was born in New York in 1882 of somewhat obscure parentage of Spanish and Irish descent. A British court sentenced him to death for his part in the Irish uprising of 1916, although this was later commuted. In 1921 he became Chancellor of the National University of Ireland; in 1932 he was made head of the government of the Irish Free State and the President of the Council of the League of Nations; and in 1938 he became President of the Assembly of that body. In 1940 he established the Dublin Institute for Advanced Studies with its Schools of Celtic Studies, Theoretical Physics and, in 1947, its School of Cosmic Physics. He was President of the Republic of Ireland between 1959 and 1973. The Royal Society of London elected him a Fellow in 1968. He died in 1975, aged nearly 93.

Erwin Schrödinger, mathematician, physicist, philosopher of science, artist, poet and linguist, inventor of wave mechanics, was born in Vienna five years after de Valera into a cultured family with some English ancestry. In the Great War he fought against the allies. He was Professor of Theoretical Physics in the University of Zürich from 1921 to 1927, and his invention of wave mechanics came in 1926. In 1927 he was called to Berlin to succeed the great Max Planck in the world's most famous professorship of

theoretical physics. In 1933 he shared the Nobel prize for physics with
P. A. M. Dirac. That year, as a declared opponent of the Nazi regime, he
was impelled to uproot himself and to leave Germany altogether.
He was invited to England, where he remained until 1936 as Fellow of
Magdalen College, Oxford. The Royal Society elected him a Foreign
Member in 1949. He was then professor of theoretical physics in Graz until
Hitler invaded Austria in 1938. After various moves, as related below, in
1940 he became the first Director (1940–45) of the School of Theoretical
Physics in the Dublin Institute, where his stay of 17 years was the longest he
ever spent in one place. He held a personal professorship in Vienna from
1956 until his death there in 1961 at the age of 73.

Fantastic were many of the turns of fortune in the careers of these two
men – none more so than the fact of their careers becoming in middle life
closely intertwined. How that came about is the main theme of this paper.
Most of the happenings now to be recounted were not in the United
Kingdom. Before proceeding to them, it must be noted from these brief
sketches of their careers that at crucial points both men owed much – in fact
their own survival – to the tolerance and generosity of individuals and
institutions in the United Kingdom. Without what this country did for
them, the rest of the story would not have been there to be told.

The plenitude of the talents that these men exercised so vigorously as
well as the turmoil of their times meant that there were many facets to their
personalities and many strands in their lives. For instance, de Valera was
unshakeable in his lifelong ideals and dreams for Ireland. He was able to
realize so many of these because they kept him alert to discern and to seize
any opportunity for furthering them. So he was at the same time an idealist
and an opportunist. All this makes it hard to present a simple story. I hope
that anything I say will be nothing but the truth, but I should despair of
ever knowing the whole truth about anything of importance for the
account.

In what has just been said one instinctively avoids calling anything about
either man 'complex'. For while each had thoughts upon innumerable
topics, his thinking upon any one topic at any one time appeared to be
essentially simple. There was something boyish about them both. One
example comes to mind. At one stage Schrödinger wrote papers on what he
called 'the final affine laws'. He submitted the first of these to the Royal
Irish Academy with a covering note to the Secretary saying, 'This paper
leaves nothing else to be done in the whole of physics – Paper II should be
ready next week', or words to that effect. The Secretary mentioned this to

de Valera, who was not much given to frivolity. But this he found to be so amusing that he recounted it to the President of the Royal Society who was visiting him at the time. And the President lost no time in telling the story on his return to the Society.

10.2. Schrödinger 1933–38

Schrödinger's Nobel Prize came in 1933. This was commemorated in Dublin on its jubilee in 1983. Apparently it was on that occasion that some participants conceived the idea for the present celebration of the centenary of Schrödinger's birth.

The year 1933 is also the natural starting time of our account. For, had it not been for what took place in Germany in 1933 Schrödinger would have been expected to retain his prestigious professorship in Berlin for something approaching the 40 years for which his predecessor Max Planck had held it. In that case de Valera's School of Theoretical Physics would probably never have been heard of.

1933 was the year when Hitler came to power with dire consequences for everybody everywhere and most immediately for intellectuals in Germany. It was through the good offices of F. A. Lindemann (Lord Cherwell) that Schrödinger was given the temporary Fellowship of Magdalen College at Oxford. Actually the award of the Nobel Prize was announced shortly thereafter. Little is on record about the next three years spent in Oxford. He seems at the time to have made little impact here in England, at any rate outside Oxford. During those years I was in London at Imperial College (making occasional visits to Oxford) yet I did not know even that Schrödinger was in the country. It seems true to say that during this interval he published no major original work in physics. His thought was evidently turning rather towards the philosophy, in a broad sense, of physical science. In 1935 he published *Science and the Human Temperament*, the first of his collections of essays in this field (Schrödinger, 1935).

According to a widely accepted account, Schrödinger then (1936) had the 'offer' of the Tait Chair of Mathematical Physics in Edinburgh, which was being vacated by its first holder C. G. Darwin, and at the same time he had an invitation to the professorship of theoretical physics in Graz (Austria). The account proceeds to assert that he acted against the advice of his friends in choosing to go to Graz. However, it is not certain that the Edinburgh proposition ever reached the stage of being a firm offer. There was an informal approach and Schrödinger visited Edinburgh where he discussed the matter with E. T. Whittaker, the head of the mathematics department. In order not to fuel premature rumours that the great man would be coming to Edinburgh, it was said that they had their talk on a walk in the Holyrood Park to avoid being seen at the University. Evidently Schrödinger did not take kindly to the idea that a professor should undertake a full share of undergraduate teaching, and he evinced no enthusiasm for the commendable Scottish tradition of the most eminent professors giving lectures to the most junior students. For such reasons, it seems, no great eagerness emerged on either side. So the attraction presented by Graz for Schrödinger to return to his native Austria, where he could enjoy what he regarded as the natural dignities of a university professor, inevitably prevailed.

The Edinburgh post went instead to another distinguished expatriate physicist, Max Born, who had been forced out of his professorship in Göttingen. There he had played a highly influential role in the creation and development of quantum mechanics and, in particular, in the interpretation of Schrödinger's wave mechanics. In passing, it has to be remarked that, while Born was successful in running a research school in Edinburgh, he left essentially all the teaching and examining of undergraduates to his assistants. He did this in just the way in which it had been apprehended Schrödinger would have done, had he been appointed! So far as the Schrödinger story is concerned, we shall see that the Whittaker–Born combination in Edinburgh came to play a crucial part. Whittaker's shrewd judgement had convinced him that Schrödinger was not the man they sought to take charge of a department in the University of Edinburgh. It was soon to convince him that Schrödinger was very much the man for the sort of Institute that de Valera dreamed of for Dublin.

10.3. de Valera and his mathematical interests

As a young man de Valera had strong academic aspirations. Owing, however, to his home economic circumstances he had a chequered career as a student in Dublin which left him in possession of no impressive qualifications on paper for an academic career. He did hold a sequence of temporary teaching appointments in academic mathematics, and he displayed ability in such work, but his attempts to secure a permanent post ultimately proved unsuccessful. His ambitions in that direction had nevertheless been far from ill-founded; those who worked with him in later times spoke of his 'scholarly instincts and habits of mind'; one said 'he was essentially an academic at heart and loved to be in the company of academics'. As he went through life he gained and retained the respect and friendship of distinguished mathematicians and mathematical physicists, among them Monsignor Patrick Browne, President of University College

Galway 1945–59, A. J. McConnell, Provost of Trinity College, Dublin 1952–74, Sir Edmund Whittaker FRS, Royal Astronomer of Ireland, 1906–12, and, of course, as we are going to see, Erwin Schrödinger, as well as many leading Irish scholars of his time. Throughout his life he maintained an active interest in mathematics, and he reportedly worked at it regularly in the evenings, consulting the afore-mentioned mathematicians about his reading. It is of interest that of the seven children of Eamon de Valera and his wife Sinead, three became university professors, one was the wife of a professor and the mother of a mathematician, two entered the legal profession, and one died in an accident as a young man. So he did very genuinely belong to the world of learning.

On the public and national side, de Valera did everything within his power to promote Ireland's academic well-being. And his gaining a position to do so much for others was a consequence of the non-fulfilment of his personal aspirations in this field! This led to his increasing involvement in republican politics, in which he rapidly rose to prominence, and so ultimately to his becoming the leader of his nation. In 1913 he joined the Irish Volunteers; in 1916 he fought as a commandant in the Easter rising, and he was the last commandant to surrender to the British, who sentenced him to death. The sentence was commuted to one of penal servitude. While serving part of it in Lewes Jail, he sent A. W. Conway what amounted to an original mathematical investigation. After being released in the general amnesty of 1917, de Valera had for several years a highly adventurous and hazardous existence, the story of which become well-known. The Irish Free State came into existence in 1922, but it was not until 1927 that de Valera and his Fianna Fail party entered Dail Eireann, the State's parliament. In 1932 the party came to power with de Valera as Head of Government and Minister for External Affairs.

The fact that it is so difficult to identify the first germ of de Valera's idea for the Dublin Institute serves to emphasise his adaptability and, in the best sense, his opportunism. For, if one version had to be abandoned, he was ready to keep on thinking of others until, when finally he arrived at one for which circumstances were propitious, he clinched the matter with tremendous determination. He was, however, steadfast in two basic ideals throughout. One was for the revitalizing of the Irish language and its associated culture. The other was for the establishment in Ireland of a centre of research in theoretical physics that should be of world renown.

The first of these had been basic to an attempt, about which de Valera had been consulted at the time, to set up an Irish National Academy. This had been discussed in 1921–22 in anticipation of the establishment of the

Irish Free State. That scheme itself proved abortive, but I think it had planted this ideal in de Valera's mind.

The other ideal seems to have developed from the hope he cherished of reviving Dunsink Observatory. This sprang from de Valera's intense admiration for Sir William Rowan Hamilton (1805–65) the great Irish mathematical physicist who had lived and worked at Dunsink from the age of 22 until his death. He believed that his government could acquire the Observatory from Trinity College Dublin, and he thought of re-opening it as a centre for work in both astronomy and geophysics. He consulted various scientists about this; Whittaker in particular, who had been at Dunsink for six years, firmly counselled against the scheme on account of the extremely poor observing conditions there. As matters were working out at the time, it was all for the best at that particular juncture for such advice to have persuaded de Valera to shelve the Dunsink proposal. We shall return to it later, but it had to be mentioned at this point because de Valera writing early in 1939 to Whittaker about his project for an 'Institute of Higher Studies' said '. . . it was in my conversation with you in regard to Dunsink, a few years ago, that the idea of this project originated . . .'. Evidently Whittaker was prepared to encourage de Valera to go for a school of theoretical physics in an institute on the model of the Institute for Advanced Studies at Princeton, instead of pursuing the Dunsink scheme. Whether it was de Valera or Whittaker who made the very first suggestion for such a school in Dublin we may never know; but it might never have been made at all unless de Valera had first put up his scheme for Dunsink. It does appear to have been de Valera's own idea, springing probably from the above-mentioned background, to include in the project a School of Irish Studies. It seems that de Valera first thought of the possibility of recruiting Schrödinger when, early in 1938, he saw a newspaper report – which may actually have been somewhat premature – that Schrödinger had lost his post in Graz following the Nazi invasion of Austria. But it does seem also that at much the same time Whittaker followed up his support for de Valera's projected school of theoretical physics by suggesting to de Valera that he might find Schrödinger to be available to go to direct its inauguration.

10.4. Schrödinger 1938–40

The Dublin Institute possesses notes made in about 1959 by Anny Schrödinger (his wife) recording some recollections of the ensuing events. I am grateful to the Institute for allowing me to base this section largely upon these notes.

Hitler invaded Austria in March, 1938. Schrödinger's friends in other countries were quick to realize his resulting plight. Several invited him to their institutions. But he was not allowed to accept, nor was he able to leave Austria.

de Valera in particular decided to invite Schrödinger to Dublin, but he concluded that it would be dangerous for Schrödinger to be known to be corresponding with anyone outside Austria. So de Valera enlisted the help of Whittaker. Whittaker consulted Max Born – by now his colleague in Edinburgh. Born wrote to Professor Richard Baer of Zürich who was a friend of both Born and Schrödinger. Baer thereupon asked a Dutch friend of his, who happened to be on his way to Vienna to try to appraise Schrödinger of de Valera's proposal. This good man called on Anny Schrödinger's mother, who lived in Vienna, and wrote on a slip of paper something like 'Mr de Valera plans to have an institute for advanced studies. Would you in principle be willing to take an appointment there?' Somehow conveying that the message was from Baer to Schrödinger he took his departure without disclosing his own identity. The mother sent the note to the Schrödingers who assimilated its intention and then destroyed it. Anny contrived a couple of weeks later to meet Richard Baer and his wife near the Swiss border to get them to send back Schrödinger's positive response to de Valera through Born and Whittaker.

In August, 1938, a curt official note informed Schrödinger that he was dismissed from his professorship in Graz. Evidently this implied a lifting of the ban upon their leaving Austria. But they concluded that it behoved them to depart as soon as they could. Presumably because of the accord between Hitler and Mussolini, Italy was the one country to which they could travel without waiting for a visa. In Anny Schrödinger's terse words, 'We left everything behind, we packed three suitcases, and three days after receiving the official letter we went off to Rome' (by train).

They got there with, one presumes, no Italian currency. One must be forgiven for repeating the account one had from a long-forgotten source about Schrödinger's handling of the situation. According to this, he got a porter to take the suitcases and to call a taxi; he gave the driver the name of an expensive hotel and told him to tip the porter; at the hotel, he told the commissionaire to settle with the driver; then he told the hotel that Enrico Fermi would settle with it. Apparently Fermi was then so famous in Rome that it all worked out happily for all concerned, and Fermi soon had the Schrödingers lodged in the Papal Academy, within the Vatican gardens, of which Schrödinger had in fact been elected a member. Apart from the fact of finishing up at the Academy this tale *may* be apocryphal, but it is entirely in character and I accepted it when I first heard it.

This happened at a time when de Valera was in Geneva as President of the Association of the League of Nations. Schrödinger wrote to him there. Only two days later the Irish Minister to the Vatican invited the Schrödingers to his legation. There de Valera spoke on the telephone to Schrödinger, and he also had the Minister procure all facilities for them to travel to England via Geneva.

In Geneva de Valera received the Schrödingers with immense kindness and consideration. The account is borne out by my own, necessarily much restricted, experience of his behaviour. de Valera had the ability to switch his whole attention from one field of interest to another (and of course back again) and to deal with each as though he never gave a thought to anything else. And obviously he cared deeply and steadfastly about everything to which he devoted his attention. At the same time he seemed never to be flustered but always to consult the convenience of the people with whom he was dealing.

The outcome of this Geneva encounter was definitely to confirm de Valera and Schrödinger in their determination to go ahead with the Dublin scheme. So de Valera's immediate concern was to get the Schrödingers to England in order to have them out of reach of the Nazis and their associates. It is as well to recall that this was about the time of the Munich negotiations between the UK and Germany.

As a first result of de Valera's efforts the Schrödingers were then able to establish a temporary base in Oxford; this being apparently aided by the fact that Schrödinger's Fellowship of Magdalen College had continued into 1938. In November of that year he made his first visit to Dublin. There he saw that it would naturally still be some time before the new institute could be in operation. So for the interim he accepted a guest professorship in Belgium sponsored by the Fondation Francqui in Brussels. He and his wife then remained in Belgium until de Valera sent for them in October, 1939. This was a month after the outbreak of World War II. By that time they were considered to be 'enemy aliens' in England, and de Valera had to negotiate a safe conduct to get them to Ireland. It must be said that over the years we have been reviewing Schrödinger did manage to display a singular talent either for taking risks or for seeming not to see that he was doing so! And his friends certainly had problems in extricating him.

In the time between leaving Oxford in 1936 and going to Dublin in 1939 Schrödinger published only two papers of much substance (on proper vibrations of the expanding universe). There seems to be no record of how he spent his time in Belgium.

So Erwin and Anny Schrödinger had arrived in the city that was to become their home for the next 17 years – longer than they ever lived in any one other place. There were to be still some months before Schrödinger could officially start work at the Institute. De Valera was able to arrange for the authorities of the Royal Irish Academy and of University College Dublin to give him temporary appointments through this interval. Even in these months of stopgap status, Schrödinger must have appreciated that he was experiencing security and tranquility such as were then available to very few of his former colleagues anywhere else in Europe.

10.5. Founding the Institute

We are thinking for the moment of the Ireland of the late 1930s. In the past the country had given to the world many famous scholars and thinkers. There had been a time in the previous century when, in particular, a good number of the world's leading mathematicians and mathematical physicists came from Ireland or worked in Ireland – among them George Boole, George Francis FitzGerald, William Rowan Hamilton, Joseph Larmor, James McCullagh, George Salmon, Gabriel Stokes and William Thomson (Lord Kelvin). Was it possible, de Valera asked, in this day and age to set up in Ireland a national scholarly institution that would ensure that once again there would be living and working in the country scholars and scientists of repute comparable to these? Could Ireland once again be a leader in some field of human intellectual endeavour?

Ireland was small and poor. Already it was beginning to appear that modern science was for the large and wealthy among the nations. However, one branch that was of the highest contemporary importance. but that required no great material resources (this was before the time of modern computers), was in fact mathematical physics. This was one subject for which Ireland had shown remarkable native talent. Fortunately for all concerned it was the field that had most attraction for de Valera himself.

Were he to persuade his country to support an institution on a worthy scale, he must be able to guarantee the coming of at least one figure of world renown. But he could not expect to secure an individual of this calibre unless he could guarantee the existence of a worthy institution.

The state of the world obviously made the situation delicate. On the one hand, it was, of course, the upheaval in the world that resulted in a man like Schrödinger being ready to contemplate a permanent move to a place like Dublin. On the other hand, the precarious state of the world made it seem a strange moment in history at which to contemplate the founding in a small country of a brand-new institution that would be nothing unless it could achieve world-wide influence. Nevertheless, where else could this be done save in a country that would not be directly involved in the coming conflict? And for a country to do this it had to be small and somewhat out of the way.

The brief quotations from de Valera that I am about to cite help to confirm how wholly clear-sighted he was about all the factors in the situation. Having reached his decision in the matter, he went ahead with determination, as well as with candour, regarding the difficulties in his way.

de Valera, let it be remembered, was Head of the Government of Ireland and Minister for External Affairs throughout 1932–48 (with later terms of office as well). Between 1939 and 1940 he served also as Minister of Education. If the action that is the concern of the present narrative seems to have taken longer than necessary, please recall that the outbreak of World War II occurred in its midst. And the man in charge of this particular action was engaged also in establishing what was to be the stance of his country in the face of this mighty conflict. With the resulting national and international matters on his hands, what is wonderful is that he could succeed at the same time in carrying through this so different enterprise.

On July 6, 1939, de Valera introduced in Dail Eireann his bill for an

'Act to make provision for the establishment and maintenance in Dublin of an Institute for Advanced Studies consisting of a School of Celtic studies and a School of Theoretical Physics, to authorize the addition to such institute of schools in other subjects and to provide for matter incidental or ancillary to the matters aforesaid.'

Speaking of the proposed School of Theoretical Physics de Valera said

'There is a branch of science in which you want no elaborate equipment, in which all you want is an adequate library, the brains and the men, and just paper. We...had in the past an important place in mathematics and theoretical physics.... This is the country of Hamilton, a country of great mathematicians. We have the opportunity now of establishing a school of Theoretical Physics...which I think will again enable us to achieve a reputation in that direction comparable to the reputation which Dublin and Ireland had in the middle of last century.... The schools will be devoted solely to the advance of learning...which will bring students of the post-graduate type from abroad.'

Later to the Senate he remarked that the time for introducing the bill might seem extraordinary, but

'if we look at it merely from the point of view of a gesture, if you like, to indicate that there is a better way than war for advancing the welfare of mankind, it may not be altogether inappropriate'.

The bill became law on June 19, 1940, its passage as already indicated having been delayed in consequence of the outbreak of war in September, 1939.

The Act provided for senior professors appointed by the President of Ireland on the advice of the government. Besides calling upon them to perform certain duties, there was thus also an honorific element in the appointment. Establishment orders under the Act set out the functions of the schools. Also they provided for the management of each school by a governing board composed of a chairman and members, likewise appointed by the President on the government's advice, along with the senior professor(s) of the school. There was also a council appointed to administer the Institute as a whole, on which the President of University College, Dublin (UCD), the Provost of Trinity College, Dublin (TCD) and the President of the Royal Irish Academy (RIA) were to serve *ex officio*. There was provision for staff members of various grades (as well as the senior professors) and also for 'scholars' who were envisaged as a rather select category of research students. Provision was made too for 'statutory lectures', to be given mainly by members of the school concerned, for the conduct of colloquia which were intended to be international in character, and for lectures by invited visiting speakers.

This original constitution has operated successfully since 1940. Here it may be noted that in 1947 there was founded, in accordance with the Act, the School of Cosmic Physics, with sections for astronomy, cosmic rays and geophysics. This was achieved with the strong support of Schrödinger and his colleagues at the time but its activities are outside the assigned scope at this contribution. It must, however, as one particularly happy feature of the whole saga, he noted how it all started from de Valera's notion of re-activating Dunsink Observatory. One very important result of the creation of the third school was that Dunsink was in fact re-opened in a way that has enabled Ireland to play some significant part in international astronomy every since.

The first two schools, Celtic Studies and Theoretical Physics, were originally located in two adjoining elegant Georgian houses in Merrion Square, close to UCD, TCD, the RIA and the National Gallery. That allocated to theoretical physics was named 'Teach Hamilton' (Hamilton House). This was the School as Schrödinger knew it; later the Institute moved to more spacious modern quarters near Ballsbridge.

In October 1940, following the procedure laid down, Erwin Schrödinger became the first senior professor in the School of Theoretical Physics. de Valera had set his heart upon having both E. T. Whittaker and A. W.

Conway also appointed to the same position at the same time. But Whittaker felt bound to agree to the request of the University of Edinburgh that he should remain as head of its mathematics department for the duration of the war. Conway had just became president-designate of UCD, and it was agreed that he would serve the well-being of Irish academic affairs even more effectively in that capacity. Also it was vitally important for this new institute that the heads of Dubin's two great university colleges should be men who were enthusiastic for its success. Both Whittaker and Conway did accept membership of the governing board. Conway in fact served as its first chairman from 1940 to 1950; Whittaker, so far as one is aware, was never able to attend a meeting, but he gave de Valera very valuable advice and encouragement.

The first chairman of the Council of the Institute was Mgr. Patrick Browne, President of University College Galway, formerly Professor of Mathematics in Maynooth College, a man of massive intellect who was another of de Valera's deeply respected advisers. The chairman of the Council was entitled to participate also in meetings of the boards; clearly he was expected to do so. Mgr. Browne was equally at home in the two original schools and the success of the entire operation owed an incalculable amount of his unbounded wisdom.

It must be clear to the reader that the present account of the Institute is concerned almost exclusively with the one school, that of theoretical physics. So in regard to the Institute as a whole, it does not do justice to the magnitude of de Valera's achievement.

10.6. Inauguration of the School of Theoretical Physics

I was professor of mathematics in the Queen's University of Belfast (QUB) 1936–44; from the time of moving to Ireland I visited Dublin several times a year and Dublin friends visited us in Belfast. Nevertheless up to 1940 I was aware of next to nothing of all the activity I have been describing.

Then there came a day in October, 1940, when quite exceptionally I happened to be in conversation with the vice-chancellor of QUB. We were interrupted by the porter coming to say that there was a telephone call for me. The vice-chancellor excused me and I took the call in the main entrance hall. A voice said, 'Mr de Valera would like to speak to you.' He came on the line, introduced himself, gave a concise description of the School of Theoretical Physics, and invited me to accept nomination to serve on its governing board. This was essentially the first I had heard about it, and yet I found myself, without hesitation, agreeing to the proposal. de Valera's few quietly spoken words had totally convinced me that this was something I ought to do.

In due course I must have told the vice-chancellor officially that I was serving on that body. But I never said anything to him about the telephone call, for I suspected him to be the sort who would have qualms about a colleague who was apt to be rung up by the head of government of a neutral state! It was an odd sort of neutrality because, of course, very many members of the British armed forces had homes in Ireland and they travelled back and forth freely.

The board held its first meeting on November 21, 1940. There were present as appointed members A. W. Conway (chairman), F. E. Hackett, A. J. McConnell, W. H. McCrea as *ex officio* member, E. Schrödinger, as chairman of the Institute, Mgr. Patrick Browne, with the registrar of the Institute in attendance.

de Valera came, and at the chairman's invitation he expounded his aims in having the School established, with some remarks about the ways in which it might operate to achieve them. He had, of course, lived with the concept for a long time, and he must have discussed it with all kinds of people. But here for the first time he was addressing those most immediately concerned in making it work. He was expressing his own ideas in his own words to a few colleagues sitting round a table. Listening to him I found it difficult to realize that he had not spent his entire life in academic circles. He showed that he knew how research is done, what conditions are conducive to getting good results, how colleagues and students should be recruited, and how the School (and the Institute as a whole) should cooperate with other academic institutions, particularly, of course, TCD and UCD. It was all done in a quiet sensitive manner, and in the fewest possible words. There was no elaboration, but one was left with the conviction that the undertaking was somehow bound to succeed.

It was at that meeting that Schrödinger was appointed the first director of the School. I suppose that was subject to endorsement by the council. We had also a proposal for the appointment of Walter Heitler to the staff. It must have been Schrödinger who originated this admirable idea. For Heitler had been a post-doctoral fellow with Schrödinger in Zürich; then he was with Born in Göttingen until he had to leave in 1933 about the same time as so many others, including Schrödinger and Born, also left Germany. Since then he had been with N. F. Mott in Bristol. Heitler came to Dublin in 1941 as an assistant professor; he became senior professor in 1943, and in 1945 he replaced Schrödinger as director, until in 1949 he was appointed to Schrödinger's former professorship in the University of Zürich.

I mention all this partly to show that from quite an early stage the office of director was treated as one that could circulate within a school.

Schrödinger resumed the directorship in 1949 and retained it until his own departure in 1956.

Here perhaps I ought to tell the rest of the story of my own rather brief active official association with the School. To the best of my recollection there were some half-dozen meetings of the governing board in its first two years. I think we discussed the appointment of a few 'scholars' and heard reports of the activities of the School from Schrödinger. Schrödinger and Heitler organized an inaugural colloquium in 1941 on meson theory, but I could not attend. In 1942 they had a somewhat more ambitious meeting lasting several days with several speakers from Ireland but also with short lecture courses given by P. A. M. Dirac and Sir Arthur Eddington. These two usually painfully shy individuals became unwontedly relaxed and sociable. All of the approximately 50 participants (myself included) from all over Ireland found the occasion a rare intellectual refreshment at a time when the war had made them feel so isolated from the scientific community elsewhere.

Early in 1943 war work took me to England and I had to tell the Institute that I should probably be unable to attend meetings again until after the war. When the war was over, I received papers for several meetings which I could not attend because the notice was so short. So I told the registrar of the Institute that I was ashamed of missing the meetings, that I was indeed anxious to attend, so would he please be so good as to let me have somewhat longer notice of dates, even if the agenda had to follow later. He solved the problem very neatly by sending me no notices at all! It was only when in due course I learned that I had been dropped from the board that I realized what had been going on. Over the years since then, I have had the pleasure of a variety of occasional contacts, but no formal association.

10.7. Schrödinger in Dublin

As has been mentioned, by the time that the School was inaugurated Schrödinger had for about a year been giving lectures in various capacities in Dublin. He was consequently already well known and admired for his talents in this regard. So far as lectures and seminars were concerned his work at the School joined up quite smoothly with what he had been doing in these other ways. Schrödinger was indeed able to start at the School with a ready-made audience.

The academic atmosphere of Dublin was congenial for Schrödinger. Its trend was to foster width of interest and the indulgence of intellectual curiosity, as against undue specialization. Of all the leading mathematical physicists – probably of all scientists of any sort – at the time, Schrödinger had much the greatest width of interest. Intellectually, Schrödinger and Dublin took to each other right from the start.

The primary object of the School was, of course, that its members should get on productively with their own research. It has often been remarked that Schrödinger almost never had a co-author in any of his own work. So the fact that for a short while at the beginning he had no colleagues in the School was no handicap in this particular respect.

At first necessarily, and afterwards quite naturally, the audiences in the School were composed of workers who were not formally members. This ensured that, right from the start, relations with the universities were excellent. The School could help the academics to do their own jobs. There was no question of it taking their jobs away, because there never was any suggestion that the Institute should be able to award degrees or diplomas. One thing the Institute did do was to provide a meeting place for workers from the two university colleges in Dublin, so that they got to know each other better than they had ever done before!

By contrast to his predilection for working alone, Schrödinger enjoyed giving public lectures, for which he had an outstanding talent. Probably his most famous performance in this field was his series 'What is Life?' given in Dublin in February, 1943. His audience numbered about 400 for every one of the several lectures.

As regards Schrödinger's success as a public speaker, this was mainly due to his sprightliness and clarity of delivery. But I always thought that, like the great M. Hercule Poirot, his conscience at times permitted him to exploit just a little quite needless brokenness in his spoken English. If he used what was not quite the standard idiom he could sometimes get his meaning across better than he could have done with the orthodox expression. The pity is that it was always the orthodox expression that appeared in any printed version.

What is Life?, subtitled *The physical aspect of the living cell* (Schrödinger, 1944), has become a famous little book going through many editions and translations. Schrödinger maintained that quantum physics is needed for the understanding of biological replication. The book has been held to have anticipated the concept of the genetic code. It may have prepared the ground for its acceptance, but it seems not to have had a direct effect upon its discovery.

The subject led Schrödinger to a philosophy of life according to which, so far as I can understand, personal existence is the sharing of universal consciousness. This he took to be the essence of ancient Indian thought. In this general context, but I do not know the exact connection, in his interpretation of quantum mechanics, it is often said that Schrödinger was nearer to Einstein than to, say, Bohr or Born. It seems to me that in such philosophical matters Schrödinger left his successors' problems to think about rather than conclusions upon which to build. He went on to some penetrating studies of the nature of space-time as well as to some clever, but perhaps less profitable, attempts to develop a unified field theory.

Returning to more mundane affairs, the day-to-day running of the School called for no significant effort on Schrödinger's part. If anything out of the ordinary occurred to worry him, I believe that he relied upon Conway to keep him right. He was always on cordial terms with de Valera, but if he considered that anything to do with the School ought to be brought to de Valera's notice, I believe that again he acted through Conway.

de Valera's interest in the Institute remained lively, and he attended special lectures and other events as often as he could.

The University of Dublin and the National University of Ireland gave Schrödinger honorary doctorates as soon as was feasible after his coming to the country. TCD made him a member of its Senior Common Room which was a rare privilege which he greatly valued; he came to feel very much at home there, lunching there frequently.

Outside the Institute and Trinity College, academic colleagues saw little of the Schrödingers in a social way. Most of these lived on the south side of the city, but the Schrödingers settled at Clontarf to the north. There they enjoyed the society of artists, literary people and people connected with the theatre. They were members of the Dublin Arts Club. Inevitably it was concluded that they had found something like the life they had known in Vienna; it was tolerant and easy-going. As seen from the other end of Dublin, Schrödinger's domestic establishment was somewhat irregular – although it was said to be entirely amicable. But he and his household were apparently made to feel accepted in Clontarf.

As a scientist Erwin Schrödinger's work in his lifetime transformed physics and chemistry and it was bidding to transform biology; as a man he is still remembered with affection and esteem in the Dublin that welcomed him in a time of adversity.

10.8. Acknowledgements

I am deeply grateful to Dr A. J. McConnell, past Provost of Trinity College, Dublin, for his help with his recollections of Erwin Schrödinger in Dublin. Also I thank Miss Eva Wills for invaluable help with material from the records of the Dublin Institute for Advanced Studies.

References

Schrödinger, E. (1935) *Science and the Human Temperament*. Allen & Unwin

Schrödinger, E. (1944) *What is Life? The Physical Aspect of the Living Cell*. Cambridge University Press

11

Do bosons condense?

J. T. LEWIS

Dublin Institute for Advanced Studies

Schrödinger's paper (1926) on Bose–Einstein condensation was submitted for publication on 15 December, 1925, immediately preceding the first of his papers on wave mechanics. Historians of science (Hanle, 1977; Klein, 1964) have seen 'an organic connection' between the two. It came between the three papers of Einstein (1924, 1925*a*, *b*) on the boson gas and Uhlenbeck's doctoral thesis (Uhlenbeck, 1927), and did not receive the attention it deserved, although, with hindsight, it is seen to contain the gem of the idea which was later to resolve the conflict between Einstein and Uhlenbeck. Einstein based his prediction (Einstein, 1925*a*) of condensation in the free boson gas on a combination of Bose–Einstein statistics and the classical density of states in phase-space; this leads to an expression for the mean particle number density as a function of kinetic energy which saturates as the total density increases so that there is a maximum possible density of particles having nonzero kinetic energy. Einstein claimed that, if the particle density is increased above this maximum, the excess goes into a state with zero kinetic energy. This is the phenomenon of condensation. Uhlenbeck (1927) objected that the result holds only 'when the quantization of translational motion is neglected'. The situation remained unclear until the Amsterdam meeting to celebrate the centenary (23 November 1937) of the birth of Johannes van der Waals; here Kramers pointed out the importance of the thermodynamic limit for the sharp manifestatioon of phase transitions. On the 23 March, 1938, Kahn and Uhlenbeck submitted a paper on the free boson gas in which Uhlenbeck withdrew his objection, conceding that Einstein's expression for the mean particle number density is correct in the thermodynamic limit (Kahn and Uhlenbeck, 1938). As we shall see, a key step in this argument is Weyl's formula for the asymptotic distribution of the eigenvalues of the Laplacian in a cube (Weyl, 1912); this formula was used by Schrödinger (1926) in his derivation of Einstein's expression.

Uhlenbeck's retraction revived interest in Bose–Einstein condensation; on 12 October, 1938, London submitted a paper in which he introduced the concept of macroscopic occupation of the ground state and discussed the coherence properties of the condensate (London, 1938). London conjectured that the superfluid phase transition in He^4 is an example of Bose–Einstein condensation. Almost fifty years later, the conjecture is still unsettled: is the effect, predicted for the free boson gas by Einstein's argument, wiped out by the interaction between the particles in liquid helium? On page 39 of his book *Superfluids*, London (1967) suggested that, as a manifestation of quantum mechanical complementarity, momentum space condensation is enhanced by a spatial repulsion among the particles. There is no rigorous proof of this. The question 'Do boson condense?' is a tantalizing one for mathematical physicists. Recently, there has been some progress on the rigorous treatment of some models of an interacting boson gas (van den Berg, Lewis and Pulè, 1987) using methods from the theory of probability. These methods have led to an increase in the conceptual understanding of existing results. In this chapter, I will illustrate this in a simple model of an interacting boson gas.

Following Gibbs, we identify the state of a system in thermal equilibrium at temperature T with the probability measure $\mathbb{P}$ on the space Ω of configurations given by

$$\mathbb{P}[A] = \sum_{\omega \in A} e^{-\beta H(\omega)} \bigg/ \sum_{\omega \in \Omega} e^{-\beta H(\omega)};$$

here $\beta = 1/kT$, where k is Boltzmann's constant, and $H(\omega)$ is the energy of the configuration ω. The normalizing factor

$$Z = \sum_{\omega \in \Omega} e^{-\beta H(\omega)}$$

is called the *canonical partition function*. For a free gas, Ω is a space of sequences of non-negative integers; the basic random variables $\{\sigma_j : j = 1, 2, \ldots\}$ are defined by

$$\sigma_j(\omega) = \omega_j$$

for each $\omega = (\omega_1, \omega_2, \ldots)$ in Ω, and σ_j is interpreted as the *occupation number of the jth energy level* $\lambda(j)$, while

$$N = \sum_{j \geqslant 1} \sigma_j$$

is interpreted as the *total particle number*. For fermion systems, ω_j is either 0 or 1; for boson systems, ω_j is unrestricted: $\omega_j = 0, 1, 2, \ldots$. In both cases we define the n-particle configuration space $\Omega_n = \{\omega : N(\omega) = n\}$ and take the full configuration space Ω to be the union of all the Ω_n, $n = 1, 2, \ldots$. For a free gas, the Hamiltonian function $H(\omega)$ is taken to be

$$H^0(\omega) = \sum_{j \geq 1} \lambda(j)\sigma_j(\omega),$$

so that, in both the fermion and boson cases, the n-particle canonical partition function $Z(n)$ is given by

$$Z(n) = \sum_{\omega \in \Omega_n} e^{-\beta H^0(\omega)}$$

and the corresponding probability measure by

$$\mathbb{P}^{(n)}[A] = Z(n)^{-1} \sum_{\omega \in A} e^{-\beta H^0(\omega)}$$

for each subset A of Ω_n.

Even for a free gas, the canonical partition function is not easy to deal with directly; instead we make use of a generating function $\Xi(\mu)$ defined by

$$\Xi(\mu) = \sum_{n \geq 0} e^{n\beta\mu} Z(n)$$

and called the *grand canonical partition function*. The corresponding grand canonical measure $\mathbb{P}^\mu$ is given by

$$\mathbb{P}^\mu[A] = \Xi(\mu)^{-1} \sum_{\omega \in A} \exp\{-\beta(H^0(\omega) - \mu N(\omega))\}$$

The advantage of doing this is that, with respect to the measure $\mathbb{P}^\mu$, the occupation numbers are independent random variables; it is straightforward to show that their distributions are geometrical:

$$\mathbb{P}^\mu[\sigma_j \geq m] = e^{m\beta(\mu - \lambda(j))}.$$

(Dependence is restored by conditioning: for A a subset of Ω_n, we have

$$\mathbb{P}^{(n)}[A] = \mathbb{P}^\mu[A \mid N(\omega) = n].)$$

The *grand canonical pressure* $p(\mu)$ is defined by

$$p(\mu) = (\beta V)^{-1} \ln \Xi(\mu),$$

where V is the volume of the system. It follows from Hölder's inequality that the function $\mu \mapsto p(\mu)$ is convex. The mean particle number density $\mathbb{E}^\mu[X]$, where $X = N/V$ and $\mathbb{E}^\mu[\cdot]$ denotes the expectation with respect to the probability measure $\mathbb{P}^\mu[\cdot]$, is given by

$$p'(\mu) = \mathbb{E}^\mu[X].$$

By varying μ, we can adjust the mean value of X.

In the fermion case,

$$\Xi(\mu) = \prod_{j \geqslant 1} (1 + e^{\beta(\mu - \lambda(j))}),$$

so that the pressure is given by

$$\overset{\circ}{p}(\mu) = (\beta V)^{-1} \sum_{j \geqslant 1} \ln (1 + e^{\beta(\mu - \lambda(j))})$$

and

$$\overset{\circ}{p}'(\mu) = V^{-1} \sum_{j \geqslant 1} \frac{1}{e^{\beta(\lambda(j) - \mu)} + 1}.$$

On the other hand, in the boson case,

$$\Xi(\mu) = \prod_{j \geqslant 1} (1 - e^{\beta(\mu - \lambda(j))})^{-1},$$

with the pressure given by

$$\overset{\circ}{p}(\mu) = (\beta V)^{-1} \sum_{j \geqslant 1} \ln (1 - e^{\beta(\mu - \lambda(j))})^{-1}$$

and

$$\overset{\circ}{p}'(\mu) = V^{-1} \sum_{j \geqslant 1} \frac{1}{e^{\beta(\lambda(j) - \mu)} - 1}.$$

It will simplify the exposition if we order the energy levels so that $\lambda(1) \leqslant \lambda(2) \leqslant \cdots$ and agree that $\lambda(1) = 0$ (there is no loss of generality in doing this; see van den Berg, Lewis and Pulè (1986a)). In the boson case, $\overset{\circ}{p}(\mu)$ exists only for $\mu < 0$; nevertheless the equation $\overset{\circ}{p}'(\mu) = \rho$ has a unique solution for each ρ in $(0, \infty)$ since $\mu \mapsto \overset{\circ}{p}'(\mu)$ is strictly increasing and tends to $+\infty$ as μ increases to zero. However, as we remarked earlier, it is necessary to go to the infinite-volume limit to get rid of finite-size effects in the expression for the pressure. As the volume increases, the energy levels become more tightly packed; for the infinite system, the energy spectrum is a continuum. To deal with this, introduce a sequence of regions Λ_l, $l = 1, 2, \ldots$, increasing with l, which eventually fills out the whole of d-dimensional Euclidean space; denote the volume of Λ_l by V_l. Associated with each region Λ_l, we have a Hamiltonian h_l which represents the energy of a free particle in the region Λ_l; the energy levels are the eigenvalues $0 = \lambda_l(1) \leqslant \lambda_l(2) \leqslant \cdots$ of h_l. Introduce the distribution function

$$F_l(\lambda) = V_l^{-1} \cdot \#\{j : \lambda_l(j) \leqslant \lambda\}.$$

Under weak assumptions on h_l and V_l (see Schrödinger (1926) for details) the *integrated density of states*

$$F(\lambda) = \lim_{l \to \infty} F_l(\lambda)$$

exists; the pressure

$$p'_l(\mu) = (\beta V_l)^{-1} \sum_{j \geqslant 1} \ln\,(1 - e^{\beta(\mu - \lambda_l(j))})^{-1}$$

$$= \int_{[0,\infty)} p(\mu|\lambda)\,\mathrm{d}F_l(\lambda),$$

where

$$p(\mu|\lambda) = \beta^{-1} \ln\,(1 - e^{\beta(\mu - \lambda)})^{-1},$$

converges, for $\mu < 0$, to $\overset{\circ}{p}(\mu)$ given by

$$\overset{\circ}{p}(\mu) = \int_{[0,\infty)} p(\mu|\lambda)\,\mathrm{d}F(\lambda)$$

and $\overset{\circ}{p}'(\mu)$ exists for $\mu < 0$ and is given by

$$\overset{\circ}{p}'(\mu) = \int_{[0,\infty)} p'(\mu|\lambda)\,\mathrm{d}F(\lambda),$$

where

$$p'(\mu|\lambda) = (e^{\beta(\lambda - \mu)} - 1)^{-1}.$$

In the standard cases, where Λ_l is a cube with Dirichlet, Neumann or periodic boundary conditions, the integrated density of states is given by $F(\lambda) = c_d \lambda^{d/2}$, the Weyl formula referred to in the introduction. For $d > 2$, $p'(0|\lambda)$ is integrable with respect to $\mathrm{d}F(\lambda)$ so that the mean particle number density $\overset{\circ}{p}'(\mu)$ cannot exceed

$$\rho_c = \int_{[0,\infty)} (e^{\beta\lambda} - 1)^{-1}\,\mathrm{d}F(\lambda).$$

This is the content of Einstein's discovery. On the other hand, the equation

$$\overset{\circ}{p}'_l(\mu) = \rho$$

has a unique solution $\mu_l(\rho)$ in $(-\infty, 0)$ for each ρ in $(0, \infty)$. This is the content of Uhlenbeck's objection. However, it can be shown (see, for example, van den Berg *et al.* (1986*a*)) that

$$\mu(\rho) = \lim_{l \to \infty} \mu_l(\rho),$$

where $\mu(\rho)$ is the unique root of

$$\overset{\circ}{p}'(\mu) = \rho, \qquad \rho < \rho_c,$$

and $\mu(\rho)$ is zero for $\rho \geqslant \rho_c$; the limit

$$\pi(\rho) = \lim_{l \to \infty} \overset{\circ}{p}_l(\mu_l(\rho))$$

exists and is given by

$$\pi(\rho) = \overset{\circ}{p}(\mu(\rho))$$

for all ρ in $(0, \infty)$. This is the equation of state. Define

$$X_l(\lambda) = V_l^{-1} \sum_{\{j \,:\, \lambda_l(j) \leqslant \lambda\}} \sigma_j,$$

the density of particles with energy less than λ; then

$$\Delta(\rho) = \lim_{\lambda \downarrow 0} \lim_{l \to \infty} \mathbb{E}_l^\rho[X_l(\lambda)],$$

where $\mathbb{E}_l^\rho[\,\cdot\,]$ denotes $\mathbb{E}_l^\mu[\,\cdot\,]$ evaluated at $\mu = \mu_l(\rho)$, exists and is given by*

$$\Delta(\rho) = (\rho - \rho_c)^+.$$

This confirms Einstein's assertion that the excess of particles, when $\rho > \rho_c$, goes into a state with zero energy. Notice that the saturation of the positive energy states arises from the minus sign in the denominator of the expression

$$\overset{\circ}{p}{}'(\mu) = \int_{[0,\infty)} \frac{\mathrm{d}F(\lambda)}{\mathrm{e}^{\beta(\lambda - \mu)} - 1}$$

for the mean particle number density and that this, in turn, comes from the boson statistics. In the fermion case, the corresponding expression is

$$\overset{\circ}{p}{}'(\mu) = \int_{[0,\infty)} \frac{\mathrm{d}F(\lambda)}{\mathrm{e}^{\beta(\lambda + \mu)} + 1}$$

and there is no saturation.

The argument we have presented is straightforward and can be made completely rigorous (van den Berg *et al.*, 1986*a*), yet it does not give much insight into why bosons condense. A more transparent argument can be given in a simple model of an interacting boson gas. This is the *mean-field model*: we define a sequence $\{H_l^a : l = 1, 2, \ldots\}$ of Hamiltonians H_l^a given by

$$H_l^a = H_l^0 + \frac{a}{2V_l} N^2$$

with $a > 0$. The term $aN^2/2V_l$, which provides a crude caricature of the interaction, can be understood classically: it arises in an 'index of refraction approximation' in which we imagine each particle to move through the system as if it were moving in a uniform optical medium and so receiving an increment of energy proportional to N/V_l; since a is positive, the interaction is repulsive. One advantage of this model is that the pressure exists in the infinite volume limit for all values of μ. The pressure $p_l(\mu)$ of the mean-field model in finite volume is given by

$$\mathrm{e}^{\beta V_l p_l(\mu)} = \sum_{\omega \in \Omega} \exp\{\beta(\mu N(\omega) - H_l^a(\omega))\}.$$

* We use $(x)^+$ to denote the positive part of x.

We can rewrite this as

$$e^{\beta V_l p_l(\mu)} = e^{\beta V_{ll}(\mu)} \mathbb{E}_l^{\mu}\left[\exp -\frac{\beta a}{2V_l} N^2\right],$$

where $\mathbb{E}_l^{\mu}[\,\cdot\,]$ is the expectation with respect to free gas probability measure. There is one drawback with this formula: this expression makes sense as it stands for $\mu < 0$ only; on the other hand, the series defining $p_l(\mu)$ converges for all values of μ in $(-\infty, \infty)$. The way around this difficulty is to fix $\alpha < 0$ and rearrange the right-hand side to give

$$e^{\beta V_l p_l(\mu)} = e^{\beta V_l \dot{p}_l(\alpha)} \mathbb{E}_l^{\alpha}\left[\exp \beta\left\{(\mu - \alpha)N - \frac{a}{2V_l} N^2\right\}\right].$$

To prove the existence of the limit

$$p(\mu) = \lim_{l \to \infty} p_l(\mu),$$

we introduce a new random variable and rewrite the expectation as an integral with respect to the distribution function of this random variable. Define $L_l(\omega, B)$ for each Borel subset of $[0, \infty)$ by

$$L_l(\omega, B) = V_l^{-1} \sum_{j \geqslant 1} \sigma_j(\omega)\delta_{\lambda_l(j)}[B],$$

where $\delta_\lambda[\mathrm{B}] = 1$ if λ is in B and is zero otherwise. Then L_l maps Ω into the space $E = M_b^+([0, \infty))$ of positive bounded measures on $[0, \infty)$. Let $\mathbb{K}_l^{\alpha} = \mathbb{P}_l^{\alpha} \circ L_l^{-1}$ be the probability measure on E; in terms of this we can write

$$\mathbb{E}_l^{\alpha}\left[\exp \beta\left\{(\mu - \alpha)N - \frac{a}{2V_l} N^2\right\}\right] = \int_E e^{\beta V_l G(m)} \mathbb{K}_l^{\alpha}(dm)$$

where

$$G(m) = (\mu - \alpha)\|m\| - \frac{a}{2}\|m\|^2$$

and

$$\|m\| = \int_{[0, \infty)} m(d\lambda).$$

If we could show that

$$\mathbb{K}_l^{\alpha}[dm] \sim e^{-\beta V_l I^{\alpha}[m]} \mu(dm)$$

for some function $I^{\alpha}[\,\cdot\,]$ and some reference measure μ then we might hope to use a Laplace-type argument to show that

$$\lim_{l \to \infty} \frac{1}{\beta V_l} \ln \int_E e^{\beta V_l G(m)} \mathbb{K}_l^{\alpha}[dm] = \sup_E (G(m) - I^{\alpha}(m)). \qquad (11.1)$$

This is not the case since there is no reference measure; nevertheless something close to it holds: the sequence $\{\mathbb{K}_l^{\alpha} : l = 1, 2, \ldots\}$ can be shown to satisfy a large deviation principle with rate-function I^α and Varadhan's theorem (Varadhan, 1966) can be applied to yield a variational formula (11.1) for the limit. The result is that the pressure $p(\mu)$ exists for all μ in $(-\infty, \infty)$ and is given by

$$p(\mu) = \inf_{m \in E} \mathscr{E}(m), \tag{11.2}$$

where

$$\mathscr{E}(m) = \int_{[0,\infty)} \lambda m(\mathrm{d}x) + \tfrac{1}{2}a\|m\|^2 - T \int_{[0,\infty)} s(\rho(\lambda))\,\mathrm{d}F(\lambda) - \mu\|m\|,$$

and

$$s(x) = k\{(1+x)\ln(1+x) - x \ln x\},$$

with

$$m(\mathrm{d}\lambda) = m_s(\mathrm{d}\lambda) + \rho(\lambda)\,\mathrm{d}F(\lambda)$$

the Lebesgue decomposition of m with respect to $\mathrm{d}F$, the density of states.

We observe that, in the expression for the functional $\mathscr{E}^\mu[\cdot]$, the term $\int_{[0,\infty)} \lambda m(\mathrm{d}\lambda)$ can be identified with the internal energy density of the free gas, the term $\int_{[0,\infty)} s(\rho(\lambda))\,\mathrm{d}F(\lambda)$ with the entropy density of the free gas, the term $\tfrac{1}{2}a\|m\|^2$ with the contribution of the interaction to the internal energy density; finally, in the term $-\mu\|m\|$, μ can be regarded as a Lagrange multiplier. In other words, the variational formula (11.2) can be regarded as giving the pressure in terms of the free-energy density $f[\cdot]$, where

$$f[m] = \int_{[0,\infty)} \lambda m(\mathrm{d}\lambda) + \tfrac{1}{2}a\|m\|^2 - T \int_{[0,\infty)} s(\rho(\lambda))\,\mathrm{d}F(\lambda);$$

notice that $f[\cdot]$ is a functional on the space of positive bounded measures on $\mathbb{R}^+$, and that the entropy density term in the functional depends only on the absolutely continuous part of the measure m.

Before discussing the solution to the variational problem (11.2), it is instructive to look at what happens if we vary first over the subset $\{m \in E : \|m\| = x\}$ of measures corresponding to a fixed particle number density x: we obtain the physically transparent formula

$$p(\mu) = \sup_{x \geq 0} \{\mu x - \overset{\circ}{f}(x) - \tfrac{1}{2}ax^2\},$$

where

$$\overset{\circ}{f}(x) = \sup_{\mu < 0} \{\mu x - \overset{\circ}{p}(\mu)\}$$

is the free-energy density of the free gas (in the notation of our earlier discussion, $\overset{\circ}{f}(x)$ can be written

$$\overset{\circ}{f}(x) = x\mu(x) - \overset{\circ}{p}(\mu(x)).)$$

Returning to the problem of solving the variational equation (11.2), we notice first that, if we make use of the Lebesgue decomposition

$$m(d\lambda) = m_s(d\lambda) + \rho(\lambda)\, dF(\lambda)$$

of a measure $m(d\lambda)$ with respect to the density of states $dF(\lambda)$ into a singular part $m_s(d\lambda)$ and an absolutely continuous part $\rho(\lambda)\, dF(\lambda)$, formal differentiation with respect to m_s and ρ yields the Euler–Lagrange equations

$$\lambda - \mu + a\|m\| = 0, \qquad\qquad \lambda \in S_m,$$

$$\lambda - \mu + a\|m\| = \beta^{-1} \ln \frac{1 + \rho(\lambda)}{\rho(\lambda)}, \qquad \lambda \in S_m^c,$$

where S_m is the support of the singular part of m. It has been proved in (van den Berg, Lewis and Pulè, 1986b) that:

(1) The infimum of $\mathscr{E}^\mu[m]$ is attained in E, the space of positive bounded measures on $\mathbb{R}^+$.

(2) Let m^* be a minimizer of $\mathscr{E}^\mu[\cdot]$; then $m^*(d\lambda) = m_s^*(d\lambda) + \rho^*(\lambda)\, dF(\lambda)$ satisfies the Euler–Lagrange equations and $\rho^*(\lambda)$ must be strictly positive on $S_{m^*}^c$ in order to maximize the entropy.

(3) If $\rho(\lambda)$ is strictly positive on S_m^c, it follows that the Euler–Lagrange equations have a unique solution with $\rho(\lambda) = \{e^{\beta(\lambda - \mu + a\|m\|)} - 1\}^{-1}$ and $a\|m\| \geqslant \mu$; hence

$$\rho(\lambda) \leqslant (e^{\beta\lambda} - 1)^{-1}$$

and the total mass of the absolutely continuous part of m cannot exceed ρ_c.

(4) Let m^* be the unique minimizer of $\mathscr{E}^\mu[\cdot]$; if $\mu > a\rho_c$ then the total mass $\|m_s^*\|$ of the singular part of m^* is strictly positive and must be concentrated at zero in order to minimize the internal energy $\int_{[0,\infty)} \lambda m(d\lambda)$ of the free gas:

$$m_s^*(d\lambda) = \|m_s^*\| \delta_0(d\lambda).$$

(5) Assume that $\|m_s^*\| \neq 0$; then it follows from (4) and the first Euler–Lagrange equation that

$$\mu = a\|m^*\| = a(\|m_s^*\| + \|m_a^*\|).$$

It then follows from the second Euler–Lagrange equation that $\rho^*(\lambda) = (e^{\beta\lambda} - 1)^{-1}$ so that $\|m_a^*\| = \rho_c$; hence $\mu > a\rho_c$. It follows that $\mu \leqslant a\rho_c$ implies $\|m_s^*\| = 0$; in this case, we find that $\rho^*(\lambda)$ is given by

$$\rho^*(\lambda) = \{e^{\beta(\lambda - \mu + a\|m^*\|)} - 1\}^{-1}$$

where $\|m^*\|$ is the unique real root of

$$x = \int_{[0,\infty)} (e^{\beta(\lambda - \mu + ax)} - 1)^{-1}\, dF(\lambda).$$

In summary: it is proved in van den Berg *et al.* (1986*b*) that there is a unique minimizer m^* of $\mathscr{E}^\mu[\,\cdot\,]$ and that m^* is given by

$$m^*(\mathrm{d}\lambda) = (\|m^*\| - \rho_c)\delta_0(\mathrm{d}\lambda) + p'(0|\lambda)\mathrm{d}F(\lambda)$$

with $\|m^*\| = \mu/a$, for $\mu > a\rho_c$, and by

$$m^*(\mathrm{d}\lambda) = p'(\mu - a\|m^*\| \,|\, \lambda)\,\mathrm{d}F(\lambda)$$

with $\|m^*\|$ the unique real root of the equation $x = \int_{[0,\infty)} p'(\mu - ax|\lambda)\,\mathrm{d}F(\lambda)$, for $\mu \leqslant a\rho_c$. Furthermore, it is shown in van den Berg *et al.* (1986*b*) that

$$\lim_{\lambda \downarrow 0}\lim_{l \to \infty} \tilde{\mathbb{E}}_l^\mu[X_l(\lambda)] = \|m_s^*\|,$$

where $\tilde{\mathbb{E}}_l^\mu[\,\cdot\,]$ is the grand canonical expectation for the mean-field model; this identifies the singular measure with the condensate. It is clear from this discussion that, in the mean-field model, bosons condense into a state of zero kinetic energy in order to minimize the free-energy once the particle number density exceeds a critical value ρ_c; below ρ_c the free-energy is minimized by having almost all particles in a state with strictly positive energy.

These methods, illustrated here in a simple example, enable us to treat more realistic models of an interacting boson gas; recent results described in my lecture make a little progress towards answering the question: 'Do bosons condense?'

References

Einstein, A. (1924) *Sitz. Preuss. Akad. Wiss.* XXII, 261–7

Einstein, A. (1925*a*) *Sitz. Preuss. Akad. Wiss.* I, 3–14

Einstein, A. (1925*b*) *Sitz. Preuss. Akad. Wiss.* III, 18–25

Hanle, P. A. (1977) *Arch. Hist. Exact Sci.* 17, 165–92

Kahn, B. and Uhlenbeck, G. E. (1938) *Physica* 5, 399–415

Klein, M. J. (1964) *The Natural Philosopher* 3, 45

London, F. (1938) *Phys. Rev.* 54, 947–54

London, F. (1967) *Superfluids*, vol. II. John Wiley and Sons, Inc., New York

Schrödinger, E. (1926) *Phys. Z.* 27, 95–101

Uhlenbeck, G. E. (1927) *Over Statistische Methoden in der Theorie der Quanta.* Thesis, Leiden

van den Berg, M., Lewis, J. T. and Pulè, J. V. (1986*a*) *Helv. Phys. Acta* 59, 1271–88

van den Berg, M., Lewis, J. T. and Pulè, J. V. (1986*b*) *On the existence of the pressure in some models of an interacting boson gas.* Preprint DIAS-STP-86-13

van den Berg, M., Lewis, J. T. and Pulè, J. V. (1987) *Commun. Math. Phys.* (in press)

Varadhan, S. R. S. (1966) *Comm. Pure Appl. Math.* 19, 261–86

Weyl, H. (1912) *Math. Ann.* 71, 441

12

Schrödinger's nonlinear optics

JAMES McCONNELL
School of Theoretical Physics, Dublin Institute for Advanced Studies

12.1. Mie–Born electromagnetic theories

In a sequence of three papers published in 1912 and 1913 Gustav Mie sought to construct a theory in which the existence of only the electromagnetic field is assumed and particles are regarded as singularities of the field (Mie, 1912a, b, 1913). He put forward the idea that the electric and magnetic susceptibilities of the vacuum are not zero. We have an electric displacement **D** which is not just a constant times the electric intensity **E** and a magnetic induction **B** which is not just a constant times the magnetic intensity **H**, as they are in Maxwell's theory. Thus the field equations are nonlinear. Mie found that there is a strong deviation from the Maxwellian results only for very great field strengths such as occur, for example, in the neighbourhood of a charged elementary particle.

Starting with a four-potential ϕ_0, ϕ_1, ϕ_2, ϕ_3, writing $\phi_4 = i\phi_0$ and defining f_{kl} by

$$f_{kl} = \frac{\partial \phi_l}{\partial x_k} - \frac{\partial \phi_k}{\partial x_l} \qquad (1.1)$$

Mie took $\mathbf{B}(f_{23}, f_{31}, f_{12})$, $\mathbf{E}(f_{14}, f_{24}, f_{34})$ as a fundamental six-vector. He then described the field by a Lagrangian density L which is a function of the six-vector and the four-potential, and he defined charge and current density as derivatives of L with respect to ϕ_k. This is unsatisfactory because it is physically impossible that field quantities should depend on the values of the potentials.

Twenty years later Max Born proposed a theory which was free of this unsatisfactory feature (Born, 1934). Shortly afterwards in collaboration with Infeld he generalized his considerations employing a relativistic formulation (Born and Infeld, 1934). In order that **E** should not diverge they postulated that there exists a maximum electric field intensity, which they denoted by b. Starting, as Mie did, with **B** and **E** we know that the Maxwell equations for free space may be deduced from

$$L = \tfrac{1}{2}(B^2 - E^2), \tag{1.2}$$

which is a relativistic invariant. Equation (1.2) may be regarded as the limiting form as b tends to ∞ of

$$L = b^2 \{ [1 + b^{-2}(B^2 - E^2)]^{1/2} - 1 \}. \tag{1.3}$$

However, there is another relativistic invariant, namely $(\mathbf{B} \cdot \mathbf{E})$, and Born and Infeld generalized (1.3) to

$$L = (1 + F - G^2)^{1/2} - 1, \tag{1.4}$$

where

$$F = \frac{B^2 - E^2}{b^2}, \qquad G = \frac{(\mathbf{B} \cdot \mathbf{E})}{b^2}. \tag{1.5}$$

Then **H** and **D** were defined by

$$\mathbf{H} = b^2 \frac{\partial L}{\partial \mathbf{B}} = \frac{\mathbf{B} - G\mathbf{E}}{(1 + F - G^2)^{1/2}} \tag{1.6}$$

$$\mathbf{D} = -b^2 \frac{\partial L}{\partial \mathbf{E}} = \frac{\mathbf{E} + G\mathbf{B}}{(1 + F - G^2)^{1/2}} \tag{1.7}$$

on employing (1.4) and (1.5). When the $(\phi_1, \phi_2, \phi_3, \phi_0)$ of (1.1) were written as (A_x, A_y, A_z, ϕ), the following field equations were deduced:

$$\mathbf{B} = \operatorname{curl} \mathbf{A}, \qquad \mathbf{E} = -\frac{1}{c}\frac{\partial \mathbf{A}}{\partial t} - \operatorname{grad} \phi \tag{1.8}$$

$$\frac{1}{c}\frac{\partial \mathbf{D}}{\partial t} = \operatorname{curl} \mathbf{H}, \qquad \operatorname{div} \mathbf{D} = 0 \tag{1.9}$$

$$-\frac{1}{c}\frac{\partial \mathbf{B}}{\partial t} = \operatorname{curl} \mathbf{E}, \qquad \operatorname{div} \mathbf{B} = 0. \tag{1.10}$$

A spherically symmetric static solution of (1.8)–(1.10) for a point charge e gives

$$D(r) = \frac{e}{r^2},$$

which diverges as $r \to 0$. On writing

$$r_0 = \left(\frac{e}{b}\right)^{1/2}, \qquad f(x) = \int_x^\infty (1 + y^4)^{-1/2} \, dy \tag{1.11}$$

it is shown that

$$\phi(r) = \frac{e}{r_0} f\left(\frac{r}{r_0}\right)$$

$$E(r) = er_0^{-2}[1 + (r/r_0)^4]^{-1/2},$$

so that $\phi(r)$ and $E(r)$ are finite as $r \to 0$. Moreover it is found for an electron with rest mass m that

$$r_0 = 1.236\frac{e^2}{mc^2} = 3.48 \times 10^{-13}\,\text{cm} \tag{1.12}$$

$$b = 3.94 \times 10^{15}\,\text{esu.} \tag{1.13}$$

In future calculations we shall choose units such that

$$e = c = b = 1 \tag{1.14}$$

and therefore, from (1.11), r_0 is equal to unity.

12.2. Schrödinger's formulation of the Born theory

About a year later there appeared a paper by Schrödinger in which results of the preceding section were expressed in terms of complex field strengths in a manner suitable for the approximate solution of the nonlinear equations (Schrödinger, 1935). He defined complex vectors $\mathfrak{F}$ and $\mathfrak{G}$ by

$$\mathfrak{F} = \mathbf{B} - i\mathbf{D}, \qquad \mathfrak{G} = \mathbf{E} + i\mathbf{H}, \tag{2.1}$$

the components of $\mathfrak{F}$ and $\mathfrak{G}$ being indicated by a suffix 1, 2 or 3. A Lagrangian density $\mathfrak{L}$ given by

$$\mathfrak{L} = \frac{\mathfrak{F}^2 - \mathfrak{G}^2}{(\mathfrak{F} \cdot \mathfrak{G})} \tag{2.2}$$

is postulated, and (1.6) and (1.7) are replaced by

$$\mathfrak{G}^* = \frac{\partial \mathfrak{L}}{\partial \mathfrak{F}} = \frac{2\mathfrak{F}}{(\mathfrak{F} \cdot \mathfrak{G})} - \frac{(\mathfrak{F}^2 - \mathfrak{G}^2)\mathfrak{G}}{(\mathfrak{F} \cdot \mathfrak{G})^2}, \tag{2.3}$$

$$\mathfrak{F}^* = \frac{\partial \mathfrak{L}}{\partial \mathfrak{G}} = -\frac{2\mathfrak{G}}{(\mathfrak{F} \cdot \mathfrak{G})} - \frac{(\mathfrak{F}^2 - \mathfrak{G}^2)\mathfrak{F}}{(\mathfrak{F} \cdot \mathfrak{G})^2}, \tag{2.4}$$

from which it immediately follows that

$$(\mathfrak{F}^* \cdot \mathfrak{G}) + (\mathfrak{G}^* \cdot \mathfrak{F}) = 0. \tag{2.5}$$

It is easily deduced from (2.2)–(2.4) that $\mathfrak{L}$ is a pure imaginary. Assuming that $\mathfrak{F}, \mathfrak{G}$ is the four-dimensional curl of a four-vector we have

$$\text{curl } \mathfrak{G} + \frac{\partial \mathfrak{F}}{\partial t} = 0, \qquad \text{div } \mathfrak{F} = 0. \tag{2.6}$$

The stress, momentum and energy tensor T_{kl} is given by

$$4\pi T_{kl} = \frac{it_{kl}}{(\mathfrak{F} \cdot \mathfrak{G})} + \frac{i}{2}\mathfrak{L}\delta_{kl}, \qquad (k, l = 1, 2, 3, 4), \tag{2.7}$$

where

$$t_{kl} = \mathfrak{F}_k \mathfrak{F}_l + \mathfrak{G}_k \mathfrak{G}_l - \tfrac{1}{2}\delta_{kl}(\mathfrak{F}^2 + \mathfrak{G}^2) \qquad (k, l = 1, 2, 3)$$

$$t_{14} = \mathfrak{G}_2 \mathfrak{F}_3 - \mathfrak{G}_3 \mathfrak{F}_2 \tag{2.8}$$

etc.

$$t_{44} = \tfrac{1}{2}(\mathfrak{F}^2 + \mathfrak{G}^2).$$

An alternative form of (2.7) and (2.8) is

$$4\pi T_{kl} = \tfrac{1}{2} \operatorname{Im} (\mathfrak{F}_l^* \mathfrak{G}_k + \mathfrak{F}_k^* \mathfrak{G}_l) - \tfrac{1}{2}\delta_{kl} \operatorname{Im} (\mathfrak{F}^* \cdot \mathfrak{G}) + \frac{i}{2} \mathfrak{L}\delta_{kl} \qquad (k, l = 1, 2, 3)$$

$$4\pi T_{14} = \operatorname{Im} (\mathfrak{F}_2^* \mathfrak{F}_3) \tag{2.9}$$

etc.

$$4\pi T_{44} = \tfrac{1}{2} \operatorname{Im} (\mathfrak{F}^* \cdot \mathfrak{G}) + \frac{i}{2} \mathfrak{L},$$

where Im denotes 'imaginary part of'.

Schrödinger introduces a Lorentz transformation that will make all the components of $\mathfrak{F}$ and $\mathfrak{G}$ vanish except the first components $\mathfrak{F}_1$ and $\mathfrak{G}_1$. He then multiplies $\mathfrak{F}_1$ by a factor $e^{i\gamma}$, where γ is real, such that $\mathfrak{F}_1$ becomes real. Thus (2.5) reduces to

$$\mathfrak{F}_1(\mathfrak{G}_1 + \mathfrak{G}_1^*) = 0. \tag{2.10}$$

For the present we exclude the possibility of $(\mathfrak{F} \cdot \mathfrak{G})$ vanishing, and so (2.10) shows that $\mathfrak{G}$ is a pure imaginary. Shrödinger writes

$$\mathfrak{G}_1 = i A \mathfrak{F}_1, \tag{2.11}$$

where A is a real constant. We see from (2.1) that the only nonvanishing members of **B**, **D**, **E**, **H** are B_1 and H_1. The field is purely magnetic. On substituting from (2.11) into (2.3) we find that

$$\mathfrak{F}_1 = \frac{(1 - A^2)^{1/2}}{A}, \qquad \mathfrak{G}_1 = i(1 - A^2)^{1/2}, \tag{2.12}$$

where A ranges between -1 and $+1$. The field defined by (2.12) is called the *standard field*. For this field (2.2) and (2.11) yield

$$\frac{i}{2} \mathfrak{L} = \frac{1 + A^2}{2A}. \tag{2.13}$$

When the fields $\mathfrak{F}$, $\mathfrak{G}$ are extremely weak, in comparison with the value of b given in (1.13), the absolute value of A in (2.12) is almost unity. Then (2.13) shows that $\mathfrak{L}$ is very close to $\pm 2i$. Hence, this is the value of $\mathfrak{L}$ for the Maxwell field.

We now examine the situation where $(\mathfrak{F} \cdot \mathfrak{G})$ vanishes. Expressing (2.3) and (2.4) as

$$\mathfrak{G}^* = \frac{2}{(\mathfrak{F} \cdot \mathfrak{G})} (\mathfrak{F} - \tfrac{1}{2}\mathfrak{L}\mathfrak{G}) \tag{2.14}$$

$$\mathfrak{F}^* = -\frac{2}{(\mathfrak{F} \cdot \mathfrak{G})} (\mathfrak{G} + \tfrac{1}{2}\mathfrak{L}\mathfrak{F}) \tag{2.15}$$

we obtain

$$(\mathfrak{F}^* \cdot \mathfrak{G}^*) = -\frac{4}{(\mathfrak{F} \cdot \mathfrak{G})} [1 + \tfrac{1}{4}\mathfrak{L}^2]. \tag{2.16}$$

If $(\mathfrak{F} \cdot \mathfrak{G})$ tends to zero, $\tfrac{1}{2}\mathfrak{L}$ tends to $\pm i$; otherwise $(\mathfrak{F}^* \cdot \mathfrak{G}^*)$ would tend to infinity and hence $(\mathfrak{F} \cdot \mathfrak{G})$ would tend to infinity, contrary to what we have supposed. Putting $\tfrac{1}{2}\mathfrak{L} = \pm i$ in (2.15) we obtain

$$\mathfrak{F}^* = -\frac{2}{(\mathfrak{F} \cdot \mathfrak{G})} (\mathfrak{G} \pm i\mathfrak{F}),$$

so that in the limit of vanishing $(\mathfrak{F} \cdot \mathfrak{G})$ we must have

$$\mathfrak{G} = \mp i\mathfrak{F}, \tag{2.17}$$

since $\mathfrak{F}^*$ is finite. On substitution into (2.1) we see that (2.17) implies

$$\mathbf{E} = \mp \mathbf{D}, \qquad \mathbf{H} = \mp \mathbf{B}. \tag{2.18}$$

Since $(\mathfrak{F} \cdot \mathfrak{G})$ vanishes, it follows from (2.17) that

$$\mathfrak{F}^2 = 0 \tag{2.19}$$

and therefore, from (2.1), that

$$B^2 - D^2 - 2i(\mathbf{B} \cdot \mathbf{D}) = 0. \tag{2.20}$$

This equation is satisfied in the limiting case of

$$E = D = H = B = 0. \tag{2.21}$$

We recall that the field quantities are expressed in units of b, so that (2.21) corresponds to the Maxwell field. Even if (2.21) is not obeyed, (2.20) may be satisfied by

$$B = D, \qquad (\mathbf{B} \cdot \mathbf{D}) = 0. \tag{2.22}$$

We then see from (2.18) that (2.22) implies that

$$E = H, \qquad (\mathbf{E} \cdot \mathbf{H}) = 0.$$

To sum up, $(\mathfrak{F} \cdot \mathfrak{G})$ vanishes for a Maxwell field or for a non-Maxwell field provided that $\mathbf{E}$ and $\mathbf{H}$ are equal in magnitude, $\mathbf{B}$ and $\mathbf{D}$ are equal in magnitude, $\mathbf{B}$ is perpendicular to $\mathbf{D}$ and $\mathbf{H}$ is perpendicular to $\mathbf{E}$.

In the Maxwell theory the signs in (2.18) are both positive. Schrödinger calls this the case of *normal fields*. Then $\mathfrak{G} = i\mathfrak{F}$ and $\mathfrak{L} = -2i$. Schrödinger considers also the case of two negative signs in (2.18), so that $\mathbf{D}$ is antiparallel to $\mathbf{E}$ and $\mathbf{B}$ is antiparallel to $\mathbf{H}$, and he refers to this as the case of *abnormal fields*. Such a duplication is present also for the standard field;

indeed we see from (2.1) and (2.11) that positive and negative values of A correspond to normal and abnormal fields, respectively. Schrödinger suggested that the normal and abnormal fields bear an analogy to the positive and negative energy solutions of Dirac's equation for the electron (Schrödinger, 1935). In subsequent calculations it will be presumed that we are concerned only with normal fields.

We deduce from (2.16) that

$$\mathfrak{L} = -2i[1 + \tfrac{1}{4}(\mathfrak{F} \cdot \mathfrak{G})(\mathfrak{F}^* \cdot \mathfrak{G}^*)]^{1/2}, \tag{2.23}$$

the minus sign being chosen so as to give the value of $-2i$ of $\mathfrak{L}$ for vanishingly small fields. On expanding the square root in (2.23) we have

$$\mathfrak{L} = -2i - \frac{i}{4}(\mathfrak{F} \cdot \mathfrak{G})(\mathfrak{F}^* \cdot \mathfrak{G}^*) + \cdots, \tag{2.24}$$

which differs from $-2i$ by a correction of fourth order in the fields. It should be noted that $\mathfrak{F} + i\mathfrak{G}$ is not a first order quantity; indeed from (2.1)

$$\mathfrak{F} + i\mathfrak{G} = \mathbf{B} - \mathbf{H} + i(\mathbf{E} - \mathbf{D}),$$

which vanishes for the Maxwell field. To find the order of $\mathfrak{F} + i\mathfrak{G}$ we deduce from (2.2) that

$$\frac{(\mathfrak{F} + i\mathfrak{G})^2}{(\mathfrak{F} \cdot \mathfrak{G})} = \frac{\mathfrak{F}^2 - \mathfrak{G}^2}{(\mathfrak{F} \cdot \mathfrak{G})} + 2i = \mathfrak{L} + 2i \tag{2.25}$$

Comparing (2.24) with (2.25) we see that $(\mathfrak{F} + i\mathfrak{G})^2$ is of sixth order and therefore that $\mathfrak{F} + i\mathfrak{G}$ is of third order.

12.3. Circularly polarized plane waves

The preceding calculations were applied by Schrödinger to the discussion of the behaviour of circularly polarized plane waves (Schrödinger, 1942a). Since we are dealing with nonlinear electromagnetic theory, the sum of field intensities belonging to each of the two waves is not a solution of the field equations of the two-wave system. A single wave is represented by the standard field quantities $\mathfrak{F}$ and $\mathfrak{G}$ defined by

$$\mathfrak{F} = C\mathfrak{a}\Omega = C\mathfrak{a} \exp\{i(\omega t - \mathfrak{t}_x x - \mathfrak{t}_y y - \mathfrak{t}_z z)\} \tag{3.1}$$

$$\mathfrak{G} = i\mathfrak{F}, \tag{3.2}$$

the A in (2.11) being taken to unity because we shall suppose the fields to be weak. C in (3.1) is a complex number whose absolute value is the amplitude, $\mathfrak{a}$ is a complex polarization vector, $|\omega|$ is the angular frequency, the value of ω being taken positive for right-handed polarization and negative for left-handed polarization. The propagation vector is $\mathfrak{t}$ multiplied by the sign of ω and we denote $|\mathfrak{t}|$ by k. We see that $k = |\omega|$, only if the velocity of propagation is unity in the units of (1.14).

For a single wave it may be shown that

$$(\mathfrak{a} \cdot \mathfrak{a}) = (\mathfrak{a}^* \cdot \mathfrak{a}^*) = 0, \qquad (\mathfrak{a} \cdot \mathfrak{a}^*) = 2$$

$$[\mathfrak{a}^*, \mathfrak{a}] = \frac{2i|\omega|}{\omega k} k, \qquad (\mathfrak{f} \cdot \mathfrak{a}) = (\mathfrak{f} \cdot \mathfrak{a}^*) = 0. \tag{3.3}$$

For two different waves with polarization vectors $\mathfrak{a}_1$, $\mathfrak{a}_2$

$$(\mathfrak{a}_1 \cdot \mathfrak{a}_2) = 2 \sin^2 \theta_{12}\, e^{i\gamma_{12}}, \qquad (\mathfrak{a}_1^* \cdot \mathfrak{a}_2^*) = 2 \sin^2 \theta_{12}\, e^{-i\gamma_{12}}$$

$$(\mathfrak{a}_1 \cdot \mathfrak{a}_2^*) = 2 \cos^2 \theta_{12}\, e^{i\gamma'_{12}}, \qquad (\mathfrak{a}_1^* \cdot \mathfrak{a}_2) = 2 \cos^2 \theta_{12}\, e^{-i\gamma'_{12}}, \tag{3.4}$$

where θ_{12} is half the angle between the propagation vectors, and γ, γ' are real numbers. It is readily found from (3.3) that

$$[\mathfrak{f}, \mathfrak{a}] = -i\frac{k\omega}{|\omega|}\mathfrak{a}, \qquad [\mathfrak{f}, \mathfrak{a}^*] = i\frac{k\omega}{|\omega|}\mathfrak{a}^* \tag{3.5}$$

and then from (3.3) and (3.5) that

$$[\mathfrak{f}, \mathfrak{b}] = \frac{ik\omega}{2|\omega|}\{(\mathfrak{b} \cdot \mathfrak{a})\mathfrak{a}^* - (\mathfrak{b} \cdot \mathfrak{a}^*)\mathfrak{a}\}. \tag{3.6}$$

For a constant vector $\mathfrak{b}$ we deduce from (3.6) that

$$\operatorname{curl}(\mathfrak{b}\Omega_1) = \frac{k_1\omega_1}{2|\omega_1|}\{(\mathfrak{b} \cdot \mathfrak{a}_1)\mathfrak{a}_1^* - (\mathfrak{b} \cdot \mathfrak{a}_1^*)\mathfrak{a}_1\}\Omega_1, \tag{3.7}$$

where the definition of Ω_1 is obvious from (3.1).

When we are considering more than one wave we express $\mathfrak{F}$ and $\mathfrak{G}$ by

$$\mathfrak{F} = \sum_l C_l \mathfrak{a}_l \Omega_l, \qquad \mathfrak{G} = i\mathfrak{F}, \tag{3.8}$$

where we suppose that the coefficients C_l are small. We shall not presume that the velocity of propagation is unity. From (3.3) and (3.8)

$$\mathfrak{F}^2 = -\mathfrak{G}^2 = 2 \sum_{m>l} (\mathfrak{a}_l \cdot \mathfrak{a}_m) C_l C_m \Omega_l \Omega_m$$

$$(\mathfrak{F} \cdot \mathfrak{G}) = i\mathfrak{F}^2, \tag{3.9}$$

so that

$$\mathfrak{L} = -2i \tag{3.10}$$

from (2.2) and (3.9). Then, from (2.3),

$$\mathfrak{F} - \tfrac{1}{2}\mathfrak{L}\mathfrak{G} = \tfrac{1}{2}(\mathfrak{F} \cdot \mathfrak{G})\mathfrak{G}^*. \tag{3.11}$$

According to (3.8) and (3.10) to the left-hand side of (3.11) vanishes while the right-hand side is equal to $\tfrac{1}{2}\mathfrak{F}^3$. So unless $\mathfrak{F}$ and $\mathfrak{G}$ vanish, (3.8) do not provide a solution of (3.11).

We therefore attempt to find a solution of (3.11) by adding to the $\mathfrak{F}$ and $\mathfrak{G}$ of (3.8) the small quantities $\mathfrak{p}$ and $\mathfrak{q}$, and writing

$$\mathfrak{F}' = \mathfrak{F} + \mathfrak{p}, \qquad \mathfrak{G}' = \mathfrak{G} + \mathfrak{q}. \tag{3.12}$$

Since $\mathfrak{p}$ and $\mathfrak{q}$ are being added in order to take account of the third order term on the right-hand side of (3.11), we presume that $\mathfrak{p}$ and $\mathfrak{q}$ are each of third order. However, as the theory is nonlinear, it is not to be expected that $\mathfrak{q}$ is equal to $i\mathfrak{p}$. On substituting from (3.12) into (3.11), employing (3.10) in the left-hand side and ignoring $\mathfrak{p}$ and $\mathfrak{q}$ on the right-hand side, as we may since we are retaining only third order terms, we obtain

$$\mathfrak{p} + i\mathfrak{q} = \tfrac{1}{2}(\mathfrak{F} \cdot \mathfrak{G})\mathfrak{G}^*$$

and so

$$\mathfrak{q} = i\mathfrak{p} - \tfrac{1}{2}i(\mathfrak{F} \cdot \mathfrak{G})\mathfrak{G}^*. \tag{3.13}$$

Substituting from (3.12) and (3.13) into

$$\operatorname{curl} \mathfrak{G}' + \frac{\partial \mathfrak{F}'}{\partial t} = 0, \qquad \operatorname{div} \mathfrak{F}' = 0$$

and noting from (3.3) that $\operatorname{div} \mathfrak{F}$ vanishes we deduce that

$$\operatorname{curl} \mathfrak{p} - i\dot{\mathfrak{p}} = \tfrac{1}{2}\operatorname{curl}\{(\mathfrak{F} \cdot \mathfrak{G})\mathfrak{G}^*\} + i(\operatorname{curl} \mathfrak{G} + \dot{\mathfrak{F}}) \tag{3.14}$$

$$\operatorname{div} \mathfrak{p} = 0. \tag{3.15}$$

At this stage we restrict the investigations to a two-wave system. We then see from (3.8) that

$$\tfrac{1}{2}(\mathfrak{F} \cdot \mathfrak{G})\mathfrak{G}^* = (\mathfrak{a}_1 \cdot \mathfrak{a}_2)C_1 C_2 \Omega_1 \Omega_2 (C_1^* \mathfrak{a}_1^* \Omega_1^* + C_2^* \mathfrak{a}_2^* \Omega_2^*)$$

$$= (\mathfrak{a}_1 \cdot \mathfrak{a}_2)C_1 C_2 (C_1^* \mathfrak{a}_1^* \Omega_2 + C_2^* \mathfrak{a}_2^* \Omega_1). \tag{3.16}$$

On substituting (3.16) into (3.14) and employing (3.3) and (3.7) we obtain

$$\operatorname{curl} \mathfrak{p} - i\dot{\mathfrak{p}} = C_1 C_2 C_2^* \frac{k_1 \omega_1}{2|\omega_1|}(\mathfrak{a}_1 \cdot \mathfrak{a}_2)\{(\mathfrak{a}_2^* \cdot \mathfrak{a}_1)\mathfrak{a}_1^* - (\mathfrak{a}_2^* \cdot \mathfrak{a}_1^*)\mathfrak{a}_1\}\Omega_1$$

$$+ \left(\frac{k_1 \omega_1}{|\omega_1|} - \omega_1\right)C_1 \mathfrak{a}_1 \Omega_1 + \overset{\longleftrightarrow}{12},$$

where $\overset{\longleftrightarrow}{12}$ signifies that we are to add terms obtained by interchanging 1 and 2.

Let us now seek a solution of (3.15) and (3.17) in the form

$$\mathfrak{p} = \mathfrak{b}_1 \Omega_1 + \mathfrak{b}_2 \Omega_2, \tag{3.18}$$

where $\mathfrak{b}_1$ and $\mathfrak{b}_2$ are time independent. On comparing (3.14) and (3.17) we notice that $i(\operatorname{curl} \mathfrak{G} + \dot{\mathfrak{F}})$ leads to

$$\left(\frac{k_1 \omega_1}{|\omega_1|} - \omega_1\right)C_1 \mathfrak{a}_1 \Omega_1 + \left(\frac{k_2 \omega_2}{|\omega_2|} - \omega_2\right)C_2 \mathfrak{a}_2 \Omega_2 \tag{3.19}$$

in (3.17). This would vanish, as $i(\operatorname{curl} \mathfrak{G} + \dot{\mathfrak{F}})$ would vanish, for the Maxwellian field. If now we choose ω_1 and ω_2 to satisfy

$$\frac{k_1\omega_1}{|\omega_1|} - \omega_1 - C_2 C_2^* \frac{k_1\omega_1}{2|\omega_1|}(\mathfrak{a}_1\cdot\mathfrak{a}_2)(\mathfrak{a}_1^*\cdot\mathfrak{a}_2^*) = 0$$

$$\frac{k_2\omega_2}{|\omega_2|} - \omega_2 - C_1 C_1^* \frac{k_2\omega_2}{2|\omega_2|}(\mathfrak{a}_2\cdot\mathfrak{a}_1)(\mathfrak{a}_2^*\cdot\mathfrak{a}_1^*) = 0, \tag{3.20}$$

we are left with

$$\text{curl } \mathfrak{p} - i\dot{\mathfrak{p}} = \tfrac{1}{2}\omega_1(\mathfrak{a}_1\cdot\mathfrak{a}_2)(\mathfrak{a}_2^*\cdot\mathfrak{a}_1)\mathfrak{a}_1^*|C_2|^2 C_1\Omega_1 + \overset{\leftrightarrow}{12}, \tag{3.21}$$

when we put $|\omega_1|$ equal to k_1 in the first term on the right-hand side of (3.17). Employing (3.7) we see that (3.21) is satisfied by

$$\mathfrak{p} = \tfrac{1}{4}(\mathfrak{a}_1\cdot\mathfrak{a}_2)(\mathfrak{a}_2^*\cdot\mathfrak{a}_1)|C_2|^2 C_1\mathfrak{a}_1^*\Omega_1 + \overset{\leftrightarrow}{12}. \tag{3.22}$$

This is the solution that we sought in (3.18). Then, from (3.13), (3.16) and (3.22),

$$\mathfrak{q} = \tfrac{1}{4}(\mathfrak{a}_1\cdot\mathfrak{a}_2)\{(\mathfrak{a}_2^*\cdot\mathfrak{a}_1)\mathfrak{a}_1^* - 4\mathfrak{a}_2^*\}|C_2|^2 C_1\Omega_1 + \overset{\leftrightarrow}{12}. \tag{3.23}$$

Equations (3.22) and (3.23) provide the third order corrections to $\mathfrak{F}$ and $\mathfrak{G}$ in (3.12).

We express (3.20) as

$$\frac{|\omega_1|}{k_1} = 1 - 2|C_2|^2\sin^4\theta_{12}$$

$$\frac{|\omega_2|}{k_2} = 1 - 2|C_1|^2\sin^4\theta_{12}, \tag{3.24}$$

by (3.4). These equations show that the velocity of propagation is not just c, which is unity in our notation, but is slightly reduced. The reduction for each wave depends on the amplitude of the other wave and on the angle between the directions of propagation of the two waves. The reduction is zero when the propagation vectors are parallel and is greatest when they are antiparallel. Independent proofs will be given in the next section of (3.24) and also of Schrödinger's result

$$w = \frac{1}{4\pi}\{|C_1|^2 + |C_2|^2 - 2\sin^4\theta_{12}|C_1|^2|C_2|^2\} \tag{3.25}$$

for the energy density w of the system of two waves, which he derived by employing (2.7)–(2.9).

We return to (3.13)–(3.15) in order to consider systems of more than two circularly polarized waves. Since at most three waves can give a contribution to the last term of (3.13), we can build up the most general pattern from combinations of two or of three waves. We denote by the subscripts 1, 2, 3 quantities related to $\tfrac{1}{2}(\mathfrak{F}\cdot\mathfrak{G})\mathfrak{G}^*$, where $\mathfrak{F}$ refers to wave 1, $\mathfrak{G}$ to wave 2 and $\mathfrak{G}^*$ to wave 3. Then we have from (3.8), (3.13)–(3.15)

$$\mathfrak{p}_{12,3} + i\mathfrak{q}_{12,3} = \{\tfrac{1}{2}(\mathfrak{F}\cdot\mathfrak{G})\mathfrak{G}^*\}_{12,3} = C_1 C_3 C_3^*(\mathfrak{a}_1\cdot\mathfrak{a}_2)\mathfrak{a}_3^*\Omega_1\Omega_2\Omega_3^* \tag{3.26}$$

$$\text{curl } p_{12,3} - i\dot{p}_{12,3} = \text{curl } \{\tfrac{1}{2}(\mathfrak{F} \cdot \mathfrak{G})\mathfrak{G}^*\}$$

$$= C_1 C_2 C_3^*(\mathfrak{a}_1 \cdot \mathfrak{a}_2) \text{ curl } (\mathfrak{a}_3^* \Omega_1 \Omega_2 \Omega_3) \quad (3.27)$$

$$\text{div } p_{12,3} = 0. \quad (3.28)$$

To solve (3.27) and (3.28) one writes

$$\omega_1 + \omega_2 - \omega_3 = \omega_4' \quad (3.29)$$

$$\mathfrak{k}_1 + \mathfrak{k}_2 - \mathfrak{k}_3 = \mathfrak{k}_4, \quad (3.30)$$

where in general $|\omega_4'| \neq \mathfrak{k}_4$. We deduce from (3.7) that

$$\text{curl } p_{12,3} - i\dot{p}_{12,3} = \frac{k_4 \omega_4'}{2|\omega_4'|} C_1 C_2 C_3^*(\mathfrak{a}_1 \cdot \mathfrak{a}_2)\{(\mathfrak{a}_3^* \cdot \mathfrak{a}_4)\mathfrak{a}_4^* - (\mathfrak{a}_3^* \cdot \mathfrak{a}_4^*)\mathfrak{a}_4\}\Omega_4'. \quad (3.31)$$

In order to obtain a solution of (3.31) we substitute

$$\Omega_4' = e^{i(\omega_4' - \omega_4)t} \Omega_4, \quad (3.32)$$

where

$$\omega_4 = \frac{k_4 \omega_4'}{|\omega_4'|}, \quad (3.33)$$

into the last term of (3.31). It is then found that

$$p_{12,3} = \frac{\omega_4}{2(\omega_4 + \omega_4')} C_1 C_2 C_3^*(\mathfrak{a}_1 \cdot \mathfrak{a}_2)(\mathfrak{a}_3^* \cdot \mathfrak{a}_4)\mathfrak{a}_4^* \Omega_4'$$

$$+ \tfrac{1}{2}\omega_4 C_1 C_2 C_3^*(\mathfrak{a}_1 \cdot \mathfrak{a}_2)(\mathfrak{a}_3^* \cdot \mathfrak{a}_4^*)\mathfrak{a}_4 \Omega_4 p(t) \quad (3.34)$$

with

$$\dot{p}(t) = -i \, e^{i(\omega_4' - \omega_4)t}. \quad (3.35)$$

The solution of (3.35) which vanishes with t is

$$p(t) = \frac{1 - e^{i(\omega_4' - \omega_4)t}}{\omega_4' - \omega_4}. \quad (3.36)$$

From the theory of radiation it is known that a significant contribution occurs only when $\omega_4' - \omega_4$ is small (McConnell, 1960). Then we have from (3.29) that $\omega_1 + \omega_2$ is approximately equal to $\omega_3 + \omega_4$. The energy density of the wave now numbered 4 being $(4\pi)^{-1}$ times the absolute square of the amplitude given by the last term of (3.34) and (3.36) as

$$\tfrac{1}{2}\omega_4 C_1 C_2 C_3^*(\mathfrak{a}_1 \cdot \mathfrak{a}_2)(\mathfrak{a}_3^* \cdot \mathfrak{a}_4^*)\mathfrak{a}_4 \Omega_4 \frac{1 - e^{i(\omega_4' - \omega_4)t}}{\omega_4' - \omega_4}$$

is

$$\frac{2\omega_4^2}{\pi} |C_1|^2 |C_2|^2 |C_3|^2 \sin^4 \theta_{12} \sin^4 \theta_{34} \frac{\sin^2 \tfrac{1}{2}(\omega_4' - \omega_4)t}{\tfrac{1}{4}(\omega_4' - \omega_4)^2}, \quad (3.37)$$

where we have employed (3.3) and (3.4).

It follows from these considerations that (3.29) and (3.30) are equivalent to

$$\omega_1 + \omega_2 = \omega_3 + \omega_4 \tag{3.38}$$

approximately, and

$$\mathfrak{k}_1 + \mathfrak{k}_2 = \mathfrak{k}_3 + \mathfrak{k}_4 \tag{3.39}$$

exactly, and we recall that the respective propagation vectors are $\mathfrak{k}_1 \omega_1 / |\omega_1|$, $\mathfrak{k}_2 \omega_2 / |\omega_2|$, $\mathfrak{k}_3 \omega_3 / |\omega_3|$, $\mathfrak{k}_4 \omega_4 / |\omega_4|$, the polarization being right-handed for positive ω and left-handed for negative ω. We may in this discussion take the velocity of propagation to be c, i.e. unity, and therefore

$$|\mathfrak{k}_i| = \pm \omega_i$$

for respective right- and left-handed polarization. It can easily be shown from (3.38) and (3.39) that, if the polarizations of waves 1 and 2 are the same (opposite), the polarizations of 3 and 4 are also the same (opposite).

12.4. Hamiltonian description of the field of circularly polarized waves

We shall first set up a classical Hamiltonian for a linear field of circularly polarized plane waves and then modify the Hamiltonian so as to account for nonlinear effects. Let us take a cube of volume V and adopt periodic boundary conditions for a system of harmonic oscillators. We denote by $\mathfrak{F}^{(s)}$ the contribution to the sth oscillator to the complex field strength $\mathfrak{F}$, we write

$$\mathfrak{F}^{(s)} = i\omega_s q_s(t) \left(\frac{4\pi}{V}\right)^{1/2} \mathfrak{a}_s \exp\{-i(\mathfrak{k}_{sx}x + \mathfrak{k}_{sy}y + \mathfrak{k}_{sz}z)\} \tag{4.1}$$

and define p_s by

$$p_s = (\dot{q}_s)^*. \tag{4.2}$$

We see from (3.1), (4.1) and (4.2) that for the unperturbed field

$$q_s = \left(\frac{V}{4\pi}\right)^{1/2} \frac{C_s}{i\omega_s} e^{i\omega_s t}, \qquad p_s = \left(\frac{V}{4\pi}\right)^{1/2} C_s^* e^{-i\omega_s t} \tag{4.3}$$

and hence that the Hamiltonian H_0 is given by

$$H_0 = i \sum_s \omega_s q_s p_s, \tag{4.4}$$

which is equal to the product of V and the energy density $\sum_s |C_s|^2/(4\pi)$.

We next set up a classical Hamiltonian H for the nonlinear fields. Since we found in the previous section that the effect of nonlinearity is to produce a third order correction to $\mathfrak{F}$, and therefore to $q_s(t)$, we shall have a fourth order correction to H_0 and so we write

$$H = i \sum_s \omega_s q_s p_s - \sum_{k \geqslant l} \sum_{m \geqslant n} A_{kl,mn} \omega_k \omega_l q_k q_l p_m p_n. \tag{4.5}$$

Thus from the canonical equations

$$\dot{q}_n = \frac{\partial H}{\partial p_n} = i\omega_n q_n - \sum_{k \geq l}\sum \sum_m A_{kl,mn}\omega_k\omega_l q_k q_l p_m. \tag{4.6}$$

In the summation term we may approximate q_k, q_l, p_m by the values in (4.3) for the linear theory, so that (4.6) becomes

$$\dot{q}_n = i\omega_n q_n + \left(\frac{V}{4\pi}\right)^{3/2} \sum_{k \geq l}\sum\sum_m A_{kl,mn} C_k C_l C_m^* \exp\{i(\omega_k + \omega_l - \omega_m)t\}. \tag{4.7}$$

We may accept (4.3) as defining q_s in the nonlinear theory, if we suppose C_s to be a slowly varying function of time. Then we have

$$\dot{q}_n = i\omega_n q_n + \left(\frac{V}{4\pi}\right)^{1/2} \frac{\dot{C}_n e^{i\omega_n t}}{i\omega_n}. \tag{4.8}$$

Now from the last term of (3.34) and from (3.35) with 1, 2, 3, 4 replaced by k, l, m, n and summed, and from (3.4) it may be shown that

$$\dot{C}_n = -2i\omega_n \sum_{k > l}\sum\sum_m C_k C_l C_m^* e^{i(\gamma_{kl} - \gamma_{mn})} \sin^2\theta_{kl} \sin^2\theta_{mn} e^{i(\omega_n' - \omega_n)t}. \tag{4.9}$$

Comparing (4.7) with (4.8) and (4.9) we see that

$$A_{kl,mn} = -\frac{8\pi}{V} e^{i(\gamma_{kl} - \gamma_{mn})} \sin^2\theta_{kl} \sin^2\theta_{mn}, \tag{4.10}$$

provided that, by (3.38) and (3.39),

$$\omega_k + \omega_l = \omega_m + \omega_n \tag{4.11}$$

is approximately true and that

$$\mathfrak{t}_k + \mathfrak{t}_l = \mathfrak{t}_m + \mathfrak{t}_m \tag{4.12}$$

is exactly obeyed; otherwise $A_{kl,mn}$ vanishes. Since $A_{kl,mn}$ vanishes anyway for $k = l$ and for $m = n$, we may by (4.10) express (4.5) as

$$H = i\sum_s \omega_s q_s p_s + \frac{8\pi}{V} \sum_{k > l}\sum\sum_{m > n} e^{i(\gamma_{kl} - \gamma_{mn})} \sin^2\theta_{kl} \sin^2\theta_{mn}\omega_k\omega_l q_k q_l p_m p_n. \tag{4.13}$$

Let us consider the case of $k = 1$, $l = 2$, $m = 1$, $n = 2$. Then (4.11) and (4.12) are obeyed and (4.13) reduces to

$$H = i\omega_1 q_1 p_1 + i\omega_2 q_2 p_2 + \frac{8\pi}{V} \sin^4\theta_{12}\omega_1\omega_2 q_1 q_2 p_1 p_2. \tag{4.14}$$

Hence, from (4.3) and (4.14),

$$\dot{q}_1 = \frac{\partial H}{\partial p_1} = i\omega_1 q_1 + \frac{8\pi}{V} \sin^4\theta_{12}\omega_1\omega_2 q_1 q_2 p_2$$

$$= i\omega_1 q_1 (1 - 2|C_2|^2 \sin^4\theta_{12}),$$

which leads to the first equation of (3.24). We also deduce from (4.14) that the energy of the two waves

$$E = \frac{V}{4\pi} \left(|C_1|^2 + |C_2|^2 - 2 \sin^4 \theta_{12} |C_1|^2 |C_2|^2 \right)$$

in agreement with the result (3.25) for the energy density of two waves.
Finally let us put $m = k$, $n = l$ in (4.13) without restricting the values of k
and l to 1 and 2. Then we shall have

$$\dot{q}_k = i\omega_k q_k + \frac{8\pi}{V} \sum_l \sin^4 \theta_{kl} \omega_k \omega_l q_k q_l p_l$$

$$= i\omega_k q_k \left(1 - 2 \sum_l |C_l|^2 \sin^4 \theta_{kl} \right).$$

This allows us to extend (3.24) to

$$\frac{|\omega_k|}{k_k} = 1 - 2 \sum_l |C_l|^2 \sin^4 \theta_{kl}. \tag{4.15}$$

It follows from (4.3) that (4.15) is expressible as

$$\frac{|\omega_k|}{k_k} = 1 - \frac{8\pi}{V} \sum_l E_l \sin^4 \theta_{kl}, \tag{4.16}$$

where E_l is the energy of the lth wave.

12.5. Scattering of light by light

We apply the foregoing theory to the mutual scattering of circularly
polarized rays. To investigate this process we employ quantum theory
starting from the classical Hamiltonian of (4.13) and interpreting the q's
and p's as operators. Thus care should be exercised in choosing the order in
which these variables are written. We must eliminate the zero-point energy,
since otherwise the quantum-mechanical analogue of (4.16) with E_l equal
to $\frac{1}{2}\hbar|\omega_l|$ would result in an infinitely strong refraction effect. The zero-point
energy may be eliminated by agreeing to write $q_s p_s$ for right-handed
polarization and $p_s q_s$ for left-handed polarization, so that

$$i\omega_s q_s p_s = n_s \hbar \omega_s \qquad \text{for right-handed polarization}$$

$$i\omega_s p_s q_s = -n_s \hbar \omega_s \qquad \text{for left-handed polarization} \tag{5.1}$$

where n_s is the occupation number for the sth level. Then (5.1) may be
satisfied by putting

$$q_s = Q_s \, e^{i\omega_s t}, \qquad p_s = P_s \, e^{-i\omega_s t}, \tag{5.2}$$

where

$$(Q_s)_{n_s+1, n_s} = \frac{[(n_s + 1)\hbar]^{1/2} \, e^{i\phi_s}}{i\omega_s^{1/2}},$$

$$(P_s)_{n_s,n_s+1} = [(n_s+1)\hbar]^{1/2}\omega_s^{1/2}\,e^{-i\phi_s} \quad (\omega_s>0) \tag{5.3}$$

$$(Q_s)_{n_s,n_s+1} = \frac{[(n_s+1)\hbar]^{1/2}i\,e^{i\phi_s}}{(-\omega_s)^{1/2}},$$

$$(P_s)_{n_s+1,n_s} = [(n_s+1)\hbar]^{1/2}(-\omega_s)^{1/2}\,e^{-i\phi_s} \quad (\omega_s<0) \tag{5.4}$$

and all other matrix elements vanish. On substitution from (5.2)–(5.4) into (4.13) we see that the perturbing Hamiltonian $H_{kl,mn}$ describing the scattering of k- and l-rays into m- and n-rays such that the k-, l-, m- and n-rays have, respectively, n_k, n_l, 0, 0 photons in the initial state and n_k-1, n_l-1, 1, 1 photons in the final state is given by

$$H_{kl,mn} = -\frac{8\pi\omega_k\omega_l}{V}\sin^2\theta_{kl}\sin^2\theta_{mn}(n_kn_l)^{1/2}\hbar^2$$

$$\times\left(\frac{\varepsilon_m\varepsilon_n\omega_m\omega_n}{\varepsilon_k\varepsilon_l\omega_k\omega_l}\right)^{1/2}e^{i(\gamma_{kl}-\gamma_{mn})}\,e^{i\alpha}\,e^{i(\omega_k+\omega_l-\omega_m-\omega_n)t}, \tag{5.5}$$

where the ε's are ±1 for positive (negative) signs of the corresponding ω's and the $e^{i\alpha}$ is a phase factor coming from the ϕ_s terms in (5.3) and (5.4).

We consider the case of the k- and l- rays being right-handed and we calculate the energy scattered within an element of solid angle $d\Omega$. Equation (5.5) reduces to

$$H_{kl,mn} = -\frac{8\pi}{V}\sin^2\theta_{kl}\sin^2\theta_{mn}\hbar^2(n_kn_l\omega_k\omega_l\omega_m\omega_n)^{1/2}$$

$$\times e^{i(\gamma_{kl}-\gamma_{mn})}\,e^{i\alpha}\,e^{i(\omega_n'-\omega_n)t}, \tag{5.6}$$

by (3.29), and we recall that (4.12) must be obeyed. From (5.6)

$$\left|\int_0^t H_{kl,mn}\,dt\right|^2 = \left(\frac{8\pi}{V}\right)^2\sin^4\theta_{kl}\sin^4\theta_{mn}\hbar^4 n_kn_l\omega_k\omega_l\omega_m\omega_n$$

$$\times\frac{\sin^2\frac{1}{2}(\omega_n'-\omega_n)t}{\frac{1}{4}(\omega_n'-\omega_n)^2}.$$

The number N of waves in $d\Omega$ is given by (McConnell, 1960)

$$N = \frac{V\omega_n^2\,d\omega_n\,d\Omega}{8\pi^3} = \frac{V\omega_n^2\,d\Omega}{8\pi^3}\frac{d(\omega_n'-\omega_n)}{2\sin^2\theta_{mn}},$$

as we can deduce from the relation

$$d\omega_m + d\omega_n' = 0.$$

The total scattered energy is found by summing $N\hbar^{-2}\left|\int_0^t H_{kl,mn}\,dt\right|^2$ over the different possible pairs of m- and n-rays. Hence the scattered energy is

$$\frac{4\,d\Omega\hbar^3}{\pi V}\int\sin^4\theta_{kl}\sin^2\theta_{mn}n_kn_l\omega_k\omega_l\omega_m\omega_n^4\times\frac{\sin^2\frac{1}{2}(\omega_n'-\omega_n)t}{\frac{1}{4}(\omega_n'-\omega_n)^2}\,d(\omega_n'-\omega_n). \tag{5.7}$$

If we disregard extremely large values of ω_n, so that the main contribution to the integral in (5.7) comes from values of ω_n such that in accordance with (4.11) energy is approximately conserved in the process, we obtain

$$\text{scattered energy} = \frac{8t\,d\Omega}{V}\,\hbar^3 \sin^4 \theta_{kl} \sin^2 \theta_{mn'} n_k n_l \omega_k \omega_l \omega_m \omega_n'^4$$

$$= \frac{8t\,d\Omega}{V}\,\hbar^3 \sin^6 \theta_{kl} n_k n_l \omega_k^2 \omega_l^2 \omega_n'^3.$$

since it may be shown from (4.11) and (4.12) that

$$\omega_m \omega_n' \sin^2 \theta_{mn'} = \omega_k \omega_l \sin^2 \theta_{kl}.$$

The incoming energy per unit area during time t is equal to $n_k \hbar \omega_k t / V$. Hence the differential cross-section for the scattering of the k-ray by one photon of the l-ray is

$$d\Phi = 8\hbar^2 \omega_k \omega_l^2 \omega_n'^3 \sin^6 \theta_{kl}\,d\Omega. \tag{5.8}$$

Expressing ω_n' as a function of scattering angle and integrating with respect to $d\Omega$ we obtain the total cross-section

$$\Phi = 16\pi\omega_k^2 \omega_l^3 (\omega_k + \omega_l) \sin^8 \theta_{kl}. \tag{5.9}$$

This result is obviously true also for two excited rays with left-handed polarization.

By choosing a suitable Lorentz frame of reference it is possible to transform the k-and l-rays so that they are in antiparallel directions, and therefore $\theta_{kl} = \frac{1}{2}\pi$, and that they have the same angular frequency ω, say. We distinguish between the following possibilities:

$\text{I}\ \omega_k < 0,\ \omega_l < 0,\ \omega_m < 0,\ \omega_n < 0;\quad \text{I}'\ \omega_k > 0,\ \omega_l > 0,\ \omega_m > 0,\ \omega_n > 0$

$\text{II}\ \omega_k < 0,\ \omega_l > 0,\ \omega_m < 0,\ \omega_n > 0;\quad \text{II}'\ \omega_k > 0,\ \omega_l < 0,\ \omega_m > 0,\ \omega_n < 0 \quad (5.10)$

$\text{III}\ \omega_k < 0,\ \omega_l > 0,\ \omega_m > 0,\ \omega_n < 0;\quad \text{III}'\ \omega_k > 0,\ \omega_l < 0,\ \omega_m < 0,\ \omega_n > 0.$

We denote by ϕ the angle between the propagation vectors of the k- and n-rays. From (5.8) and (5.9) we deduce that

$$d\Phi = 8\hbar^2 \omega^6\,d\Omega \tag{5.11}$$

I and I'

$$\Phi = 32\pi\hbar^2 \omega^6. \tag{5.12}$$

For the other combinations of polarization in (5.10) it may similarly be shown that

$$d\Phi = 8\hbar^2 \sin^6 \tfrac{1}{2}\phi\omega^6\,d\Omega \tag{5.13}$$

II and II'

$$\Phi = \tfrac{32}{5}\pi\hbar^2 \omega^6 \tag{5.14}$$

$$d\Phi = 8h^2 \cos^8 \tfrac{1}{2}\phi \omega^6 \, d\Omega \qquad (5.15)$$

III and III′

$$\Phi = \tfrac{32}{5}\pi h^2 \omega^6. \qquad (5.16)$$

We see from (5.11), (5.13) and (5.15) that the differential cross-section is independent of the scattering angle only if the two excited rays have both right-handed or both left-handed polarizations. We see from (5.15) that, when the polarizations of the initial rays are opposite, there is a strong tendency to conserve the polarization. We see from (5.12), (5.14) and (5.16) that the total cross-section has the same value whenever the polarizations of the initial rays are opposite, and that it is only one-fifth of the value when the polarizations are the same.

12.6. Linearly polarized plane waves

We shall briefly investigate the mutual scattering of linearly polarized rays. This may be done by a canonical transformation of variables (McConnell, 1943)

$$2^{1/2} i\omega q_1 = p_R + i\omega q_L, \qquad 2^{1/2} p_1 = p_L - i\omega q_R ,$$
$$2^{1/2} i\omega q_2 = p_R - i\omega q_L, \qquad 2^{1/2} p_2 = -p_L - i\omega q_R , \qquad (6.1)$$

In (6.1) the suffixes 1, 2 refer to the linear polarizations, the R and L suffixes indicate right- and left-handed circular polarization and ω now denotes the angular frequency taken positive for both R and L. If the interaction between the waves is neglected,

$$(q_1)_{n,n+1} = i\left[\frac{(n+1)\hbar}{\omega}\right]^{1/2} e^{i(\omega t + \phi_1)}, \quad (p_1)_{n+1,n} = [(n+1)\hbar\omega]^{1/2} e^{-i(\omega t + \phi_1)} \qquad (6.2)$$

$$(q_2)_{n,n+1} = i\left[\frac{(n+1)\hbar}{\omega}\right]^{1/2} e^{i(\omega t + \phi_2)}, \quad (p_2)_{n+1,n} = [(n+1)\hbar\omega]^{1/2} e^{-i(\omega t + \phi_2)},$$

where $e^{i\phi_1} \cdots e^{-i\phi_2}$ are arbitrary phase factors.

We consider the scattering of a pair of antiparallel rays of equal frequency into another pair of antiparallel rays. We now denote by $H_{kl,mn}$ the part of the perturbing Hamiltonian which gives the scattering of the k- and l-rays into the m- and n-rays. It may be proved that $H_{kl,mn}$ will vanish unless $klmn$ is one of the combinations

$$1111 \qquad 1122 \qquad 1221 \qquad 1212 \qquad (6.3)$$

$$2222 \qquad 2211 \qquad 2112 \qquad 2121 \qquad (6.4)$$

and that the results for (6.4) will be the same as those for the respective members of (6.3). On performing the calculations it is found that

$$H_{kl,mn} = -\frac{2\pi}{V} h^2 \omega^2 q_k q_l p_m p_n (y(\phi))^{1/2}, \qquad (6.5)$$

where for the combinations (6.3)

$$
\begin{array}{ll}
1111 & y(\phi) = (3 + \cos\phi)^2 \\
1122 & y(\phi) = \sin^4\phi \\
1221 & y(\phi) = 4(1 + \cos\phi)^2 \\
1212 & y(\phi) = 4(1 - \cos\phi)^2.
\end{array}
\tag{6.6}
$$

Then from (6.2) and (6.5) with the phase factors omitted, as they may be for the calculations of scattering cross-sections,

$$
H_{kl,mn} = -\frac{2\pi}{V}\hbar^2(n_k n_l)^{1/2}\omega^2(y(\phi))\,e^{i(\omega_m \omega_n - \omega_k - \omega_l)t},
\tag{6.7}
$$

which leads to the differential cross-section

$$
d\Phi = \tfrac{1}{2}\hbar^2\omega^6 y(\phi)\,d\Omega.
\tag{6.8}
$$

The total scattering cross-sections for the combinations (6.3) are given by

$$
\Phi_{1111} = \frac{28\pi}{5}\hbar^2\omega^6
$$

$$
\Phi_{1122} = \frac{4\pi}{15}\hbar^2\omega^6
\tag{6.9}
$$

$$
\Phi_{1221} = \Phi_{1212} = \frac{8\pi}{3}\hbar^2\omega^6.
$$

From (6.6) and (6.9) we see that, if two excited rays have like polarizations, the produced rays have also like polarizations, and their polarizations are very probably the same as those of the excited rays. If the excited rays have unlike polarizations, the produced rays have also unlike polarizations.

In order to compare our results with those of other investigators we express them in cgs units. We recall from (1.12) and (1.14) that we chose units such that

$$
e = c = b = r_0 = 1.236\frac{e^2}{mc^2} = 1,
$$

where m is the rest mass of the electron. We have to multiply the $\hbar^2\omega^6$ in (6.8) and (6.9) by a factor that will produce a quantity of dimension (length)^2. We shall also express results in terms of wavelength λ rather than angular frequency, employing the relation $\lambda\omega = 2\pi c$. Thus we obtain

$$
\hbar^2\omega^6 = \frac{2^6\pi^6 r_0^8}{\alpha^2\lambda^6},
$$

where α is the fine-structure constant $e^2/\hbar c$. Hence (6.8) transforms to

$$
d\Phi = \frac{2^5\pi^6 r_0^8}{\alpha^2\lambda^6}y(\phi)\,d\Omega.
\tag{6.10}
$$

For the 1111 combination of polarizations we deduce from (6.6), (6.9) and (6.10)

$$d\Phi_{1111} = \frac{2^5 \pi^6 r_0^8}{\alpha^2 \lambda^6} (3 + \cos^2 \phi)^2 \, d\Omega \tag{6.11}$$

$$\Phi_{1111} = \frac{2^{10} \times 7\pi^7 r_0^8}{5\alpha^2 \lambda^6}. \tag{6.12}$$

On substituting 137 for α^{-1} in (6.11) and (6.12) we obtain

$$d\Phi_{1111} = 5.77 \times 10^8 r_0^8 \lambda^{-6} (3 + \cos^2 \phi)^2 \, d\Omega \tag{6.13}$$

$$\Phi_{1111} = 8.12 \times 10^{10} r_0^8 \lambda^{-6}. \tag{6.14}$$

Hans Euler investigated the same problem by treating the process as occurring through the virtual production and annihilation of electron–positron pairs in intermediate states (Euler, 1936). His result for the differential cross-section is expressible as

$$d\Phi_{1111} = \frac{2^5 \pi^4 r_0^8 (3 + \cos^2 \phi)^2}{3^4 \times 5^2 \times 1.236^8 \lambda^6 \alpha^4} \, d\Omega. \tag{6.15}$$

It is interesting to note that the dependences on ϕ, on r_0 and on λ are the same in (6.11) to (6.15). On substitution of the value of α it is found from (6.15) that

$$d\Phi_{1111} = 10^8 r_0^8 \lambda^{-6} (3 + \cos^2 \phi)^2 \, d\Omega, \tag{6.16}$$

from which it follows that

$$\Phi_{1111} = 1.41 \times 10^{10} r_0^8 \lambda^{-6}. \tag{6.17}$$

Comparing (6.16) and (6.17) with (6.13) and (6.14) we see that the rate of scattering deduced from the Born theory is about six times that coming from the Dirac theory. That the scattering rates are of comparable magnitude is remarkable, since the Dirac and Born theories have very little in common. We may however recall from section 14.2 that one year before the publication of Euler's paper Schrödinger had suggested that an analogy exists between Born's normal and abnormal fields and the positive and negative energy solutions of Dirac's equation.

12.7. Conclusion

We have described only one of the several applications of the Born theory, which Schrödinger so elegantly presented and developed. The paper on nonlinear optics also contains a discussion of the scattering of light by the strong fields that are found in the neighbourhood of a point charge. Then he devoted a lengthy paper to the discussion of the dynamics and scattering power of the electron in Born's theory (Schrödinger, 1942b). Later he reintroduced consideration of abnormal fields in order to find an exact solution for a system of two waves (Schrödinger, 1943).

Schrödinger pointed out more than once it was not to be hoped that his calculations would lead to detectible experimental results. This could be expected from the value 3.94×10^{15} esu of the field b, which serves as the unit of field intensity. It was also seen from Schrödinger's study of black-body radiation, where it was shown that the Planck law for the specific intensity at a given temperature and frequency would undergo a correction of 1% only at temperatures exceeding 10^{10} K.

References

Born, M. (1934) *Proc. R. Soc. A* 143, 410–37

Born, M. and Infeld, L. (1934) *Proc. R. Soc. A* 144, 425–51

Euler, H. (1936) *Ann. Phys.* 26, 398–448

McConnell, J. (1943) *Proc. R. Ir. Acad. A* 49, 149–76

McConnell, J. (1960) *Quantum Particle Dynamics*, 2nd edn. North-Holland Publishing Co., Amsterdam

Mie, G. (1912a) *Ann. Phys.* 37, 511–34

Mie, G. (1912b) *Ann. Phys.* 39, 1–40

Mie, G. (1913) *Ann. Phys.* 40, 1–65

Schrödinger, E. (1935) *Proc. R. Soc. A.* 150, 465–77

Schrödinger, E. (1942a) *Proc. R. Ir. Acad. A* 47, 77–117

Schrödinger, E. (1942b) *Proc. R. Ir. Acad. A* 48, 91–122

Schrödinger, E. (1943) *Proc. R. Ir. Acad. A* 49, 59–66

13

Schrödinger's unified field theory seen 40 years later

O. HITTMAIR

Technische Universität Wien

13.1. The prehistory to Schrödinger's activity in unifed field theory

Unification is one of those long-standing quests of science. A superior point of view allows one to recognize connections and to uncover common roots. In the theory of general relativity Einstein succeeded in achieving a superior point of view in an especially impressive manner. By extension of the theory of special relativity he was able to comprehend gravitation in the geometrization of the space-time-continuum. After the success of this process of geometrization, the inclusion and unification of the electromagnetic field and possibly other fields could be considered to be a particularly important goal.

A few years after the discovery of general relativity, Weyl (1918) had already tried a fundamental extension of the framework of the theory in order to include electromagnetism as well. His attempt was based on the idea of gauge invariance – a concept which was to emerge 30 years later as a cornerstone of the modern theories of unification. Nevertheless, no agreement with observed facts could be reached by his concept of the path-dependence of a displaced length.

Whereas Weyl's generalization consisted in placing a connection of lengths beside the connection of directions given by the metric, Eddington (1923) followed a different course in considering the connection of directions as an *a priori* property of the manifold. In this case the metric becomes a deduced quantity. An action principle with suitably chosen Lagrangian, which contains the components of the affine connection in an appropriate form, should lead to the equations for the gravitational and the electromagnetic fields.

Eddington, and later Einstein (1923), tried to make progress in this direction. They suspended their attempts, however, on account of the multitude of possibilities which had not seemed to permit an advance free of arbitrariness. Both of them, however, continued their endeavors towards a unified field theory in different directions. Today, one is inclined to say that their critical attitude towards the affine theory was justified.

13.2. Schrödinger's approach and the nature of the envisaged field theory

In 1943 Schrödinger followed the exact path taken by Eddington and Einstein in order to arrive at a unified field theory; i.e. he considered the components of the affine connection as independent variables. The field equations were derived from an action principle, in whose Lagrangian these components entered through the contracted curvature tensor, the Ricci tensor, without first specifying how the Lagrangian contains this tensor. This fact alone furnishes the general framework of the field equations, which now comprise the gravitational field, the electromagnetic field and the meson field which should represent the nuclear forces. The two points which Schrödinger claimed for the progress he achieved are:

(1) the proper manipulation of the Lagrangian so that for the present its particular form need not be directly specified, and

(2) the introduction of the third field, the meson field, in the unification.

To construct the final form of the field equations the explicit form of the Lagrangian must be known. In his later papers Schrödinger even claimed to have taken the only justifiable choice. However, there is no longer any talk of the meson field, only of a certain coincidence with later attempts of Einstein to unify the gravitational and the electromagnetic fields.

The aims of Schrödinger do correspond largely to those of Einstein, namely to construct a unified field theory of purely classical fields into which no external matter is introduced and where the particles originate from the field properties.

Also, Schrödinger's meson field was to be understood as a classical field, the classical analogue of the quantized field of nuclear forces. In his opinion one has to understand the classical fields before proceeding to their quantization. Einstein, on the other hand, hoped to get along without any quantization at all.

With respect to this macroscopic field of nuclear forces, Schrödinger mentioned that the field equations in linear approximation amount to the Proca equations for vector mesons. Therefore in this interpretation he made use of the microscopic theory, without, however, considering physical variables like spin or isospin. This is remarkable in view of the fact that Heitler was working in the same institute at the same time on the $SU(2)$

group. Also, the extensions to the general theory of relativity by E. Cartan were not taken into account by Schrödinger, although at that time they had long since been known. The third field introduced by Schrödinger should act as a compensator for difficulties arising in the unification of the other two. So, the hope is pronounced that gauge invariance, the loss of which inevitably occurred in Schrödinger's nonlinear electrodynamics of the unified fields, could be reinstated by the addition of the third field. Gauge invariance, however, is considered to be an embellishment rather than the fundamental trait of the theory in modern unification.

13.3. The special description of the electromagnetic field

The equations for the electromagnetic field which Schrödinger obtained from his action principle are only approximately the Maxwell equations. They are the equations of the Born–Infeld description of the field (Born and Infeld, 1934), which in turn are based on Mie's theory (Mie, 1912a, b, 1913). This theory tries to explain the existence of electric charges as inner properties of the field by means of the altered field equations. Outside the charged particles the equations turn into Maxwell's equations.

For this purpose field equations for a vacuum were set up which were formally identical with the ones in matter, i.e. they contained two antisymmetric tensors ('six-vectors') of which one, the 'matter six-vector', was only an auxiliary quantity and in itself was a function of the original six-vector for the vacuum and of the four-potential. The same is true for the four-current density. These functions can be obtained from an action principle. They require new interpretation of the four-potential which follows from the choice of the Lagrangian. The equations are of course nonlinear.

Born and Infeld had avoided this difficulty of a new interpretation of potentials by connecting the components of the auxiliary six-vector as partial derivatives of the Lagrangian with the components of the original vacuum six-vector. In this way, corresponding equations result as a first step without specifying the Lagrangian. By making this choice, the special functional connection (and with it the meaning of the auxiliary six-vector) is established. In this way, a finite self-energy of an electric point charge is achieved. Attempts to quantize the theory failed. Conceptually, these efforts were formally analogous to exactly the direction pursued by Schrödinger with his unified field theory. One supposes that the occupation with the work of Born and Infeld to which Schrödinger had devoted himself for several years promoted his turning to the unified field theory. The methodological analogy between his efforts in this field and the approach of Born and Infeld is apparent.

13.4. A new attempt is started

In 1943 Schrödinger presented a paper to the Royal Irish Academy on which he later bestowed the acronym GUT. It has the title 'The general unitary theory of the physical fields' (Schrödinger, 1943a), and now it has more or less fallen into oblivion. The acronym, however, has been reintroduced for the modern grand unification theory of the gauge group $SU(3) \otimes SU_L(2) \otimes U(1)$.

GUT was for Schrödinger more than an abbreviation; he was fascinated by his results, and, although he had to change some statements in later work, GUT was always for him typical for the whole line of work in objective and execution.

In a letter to a friend he wrote:

'I have found the unitary field equations. They are based only on primitive affine geometry, a way which Weyl opened and Eddington extended, whereupon Albert did the main job in 1923, but missed the goal by a hair's breadth. The result is fascinatingly beautiful. I could not sleep for a fortnight without dreaming of it.'

So Schrödinger took up the ideas of Weyl, Eddington and Einstein to start from the components of the affine connection and to form a Lagrangian of them. His new input was that this Lagrangian should depend on these components only by way of the Ricci tensor. Then the action principle already furnished the field equations in general form. In a further step the special form of the Lagrangian was assumed. This yielded the mutual relations of the quantities appearing in the field equations, i.e. the physical content of the theory.

13.5. The general framework

The affine connection results from the fact that the components A^k of a four-vector change in a parallel displacement dx^l by

$$\delta A^k = -\Gamma^k_{ml} A^m \, dx^l \tag{5.1}$$

The Lagrangian $\mathcal{L}$, whose variation by the Γ^k_{ml} should vanish, contains only the Ricci tensor

$$R_{kl} = -\frac{\partial \Gamma^\alpha_{kl}}{\partial x^\alpha} + \Gamma^\alpha_{k\beta} \Gamma^\beta_{\alpha l} + \frac{\partial \Gamma^\alpha_{k\alpha}}{\partial \chi^l} - \Gamma^\alpha_{kl} \Gamma^\beta_{\alpha\beta} \tag{5.2}$$

$$\mathcal{L} = \mathcal{L}(R_{kl}) \tag{5.3}$$

$$\delta \int \mathscr{L} \, d\tau = \int \frac{\partial \mathscr{L}}{\partial R_{kl}} \, \delta R_{kl} \, d\tau = 0 \qquad (5.4)$$

Equation (5.4) yields the initial form of the field equations, the partial derivatives $\partial \mathscr{L}/\partial R_{kl}$ furnish the basis for a metric and field quantities. For the interpretation of these quantities the special form of $\mathscr{L}$ and its Legendre transform are necessary.

For the metric density

$$\mathfrak{g}^{kl} = \sqrt{(-g)} g^{kl} = \frac{1}{2} \left(\frac{\partial \mathscr{L}}{\partial R_{kl}} + \frac{\partial \mathscr{L}}{\partial R_{lk}} \right) \qquad (5.5)$$

is chosen. Therefore the Γ^m_{kl} by way of the R_{kl} determine the metric g^{kl}. Its determinant is g. Some comments may be made here:

(1) Schrödinger (1943a) assumed $\Gamma^k_{lm} = \Gamma^k_{ml}$, although he soon withdrew this constraint;
(2) nevertheless $R_{kl} \neq R_{lk}$ holds, which can be traced back to the third term on the r.h.s. of (5.2). This fact allows both the electromagnetic and the meson fields to be considered. In a purely gravitational field this term is symmetric in k and l and the Γ^m_{kl} are the Christoffel brackets;
(3) the method has a forerunner in Mie's electrodynamics (Mie 1912a, b, 1913), where Mie, by additional fields, explains charges and currents. In doing so he also accomplishes something like a unification, since the new electromagnetic field becomes the only physical quantity. Also in this case a general frame of provisional field equations results from an action principle, whereupon, by the special form of the Lagrangian, the final set of nonlinear field equations is obtained which coincides with the Maxwell equations in the region unoccupied by charges. A certain explicit form of the Lagrangian leads to the equations of Born and Infeld (1934) which Schrödinger had studied intensively in the preceding years.

13.6. The special execution

A special form for $\mathscr{L}$ is assumed. This leads to a system of coupled equations linking the so-called matter tensor of the gravitational field with the electromagnetic energy tensor of the Born–Infeld theory. It also contains equations for the electromagnetic field which are, in Schrödinger's terminology, of Born–Infeld character. In reality they are of the Mie-type, since four-current density and four-potential are proportional to each other. This only makes sense with a new interpretation of the potentials. Schrödinger, however, was not disturbed by this at the time. On the contrary, in the same manner he added another field, the meson field.

13.7. Tentative inclusion of nuclear forces

In GUT Schrödinger took the view that at each point of space-time another
affine connection of symmetric type may be assumed as long as it yields the
same metric which must be unique on account of length and time
measurements. By this 'duplication' it is possible to accommodate the
meson field for the nuclear forces, in the form of a classical field of very
short range. This is achieved by adjusting a constant in the Lagrangian.

   Such a duplication must be considered a weak point of the theory
because of its inherent ambiguity. But a friend of Schrödinger, A. J.
McConnell of Trinity College Dublin, immediately after reading the GUT
paper, noted that instead of this strange duplication of a symmetric affine
connection, one could use a non-symmetric one right from the beginning.
This idea was taken up by Schrödinger at once and he relied from then on
only on a nonsymmetric affine connection. The symmetric part of the affine
connection, which was treated as above, turned out to represent the
gravitation and meson fields. The additional field which arises from
dropping the symmetry condition in Schrödinger's revised version is
necessarily identified with the electromagnetic field.

13.8. Efforts towards a confirmation by measurement

It would be wrong to assume that Schrödinger had contented himself with
the derivation and interpretation of field equations in the proud view that
only fundamental relations would be worthy of his efforts. It is impressing
to see how eagerly he followed up the geophysical consequences of his
particular equations for the electromagnetic field (Schrödinger, 1943b). Are
there observations, he asked himself, which could only be explained by the
property of the four-current to be proportional to the four-'potential'?

   If gravitation is neglected, the equations for a weak electromagnetic field
according to GUT become $(c = 1)$

$$\mathbf{H} = \text{curl } \mathbf{A}$$

$$\mathbf{E} = -\frac{\partial \mathbf{A}}{\partial t} - \text{grad } \varphi$$

$$\text{curl } \mathbf{H} - \frac{\partial \mathbf{E}}{\partial t} = -\mu^2 \mathbf{A} \tag{8.1}$$

$$\text{div } \mathbf{E} = -\mu^2 \varphi$$

where the value of μ must be small enough to lead to the Maxwell equations
on a terrestrial scale, but, on the other hand, μ must be large enough so that
it becomes effective in the explanation of the permanent magnetic fields of

the Earth and of the Sun. To Schrödinger this appeared to be quite possible and he quoted extensive material of geophysical measurements which should have confirmed the explanatory value of (8.1) for the permanent magnetic field of celestial bodies.

But, as already mentioned, in the very same year Schrödinger dropped the symmetry condition of the Γ_{kl}^m and changed horses.

13.9. The unification of the three fundamental fields

Now, the field equations were derived in close analogy to GUT, but with an antisymmetric part of Γ_{kl}^m, called torsion (Schrödinger, 1944). For the general set of equations there is no specification of the Lagrangian. But already here one realizes that the new equations stemming from the torsion can only represent the electromagnetic field. Therefore the field equations obtained in GUT represent gravitation and the meson field if the three fields, gravitation, meson and light, are taken as the only possible and complete set of real fields in nature.

At first, a limited asymmetry of Γ_{kl}^m, a so-called intermediate symmetry condition, was imposed, but then it was seen that this need not be maintained and the full asymmetry with all 64 components of the affine connection was adopted (Schrödinger, 1946).

Only now is the full generality of the theory achieved and the field quantities obtained are 'in admirable agreement with the requirements of physics', which are given by the three then known fields. It is considered as a confirmation that the linear approximation of the meson equations assume the form of the Proca equations for the vector meson.

Another satisfying point is that, in the theory of the three fields, gauge invariance of the Γ_{kl}^m is restored, which, as mentioned above, had been lost in GUT. This, however, is not considered a matter of great importance.

Schrödinger was of the opinion that he had now finally found the proper affine field theory, since it was based on the general affine connection in space-time. The only further condition was that in the provisional derivation of field equations only the two contractions of rank two of the curvature tensor – one of them the Ricci tensor – were implied. The special form of $\mathscr{L}$ was not needed at the point. He wrote 'Whether it is physically right or wrong it must be called the [very] affine field theory' (Schrödinger, 1946).

Schrödinger surely was right in his opinion on this affine field theory. Only it could not be successful as such. Too much arbitrariness was involved in the choice of the Lagrangian and the field quantities. Besides, it appears very doubtful whether a macroscopic theory of nuclear forces which was constructed according to Mie's theory could provide useful information for experimental observation. The meson field is 'violently nonlinear' and its linear approximation (Proca equations) surely is not valid in the nuclear region. Schrödinger himself already expressed doubts whether the classical equations obtained for the meson field were of particular usefulness for finding the corresponding quantum relations.

13.10. A last culmination with the 'right' Lagrangian

With the especially simple Lagrangian

$$\mathcal{L} = \frac{2}{\lambda} (-\det R_{\mu\nu})^{1/2} \tag{10.7}$$

Schrödinger experienced once more a culmination of his work in unified field theory (Schrödinger, 1947). He wrote 'Now the correct Lagrangian is found, the fog sinks and everything becomes much simpler'. Again the Lagrangian (10.7) only depends on the Ricci tensor. Another dependence (Schrödinger, 1946) is not taken into account. Correspondingly, the resulting equations only cover two fields, namely gravitation and electromagnetism. λ is a constant which appears in the field equations as a cosmological constant. In contrast to (5.5) the division into symmetric and antisymmetric parts now pertains to the metric-density itself:

$$\mathfrak{g}_{ik} = \frac{\partial \mathcal{L}}{\partial R_{ik}}. \tag{10.8}$$

Using the so-called star connection

$$^*\Gamma^i_{kl} = \Gamma^i_{kl} + \tfrac{2}{3}\delta^i_k\Gamma_1 \tag{10.9}$$

with

$$\Gamma_l = \tfrac{1}{2}(\Gamma^\sigma_{l\sigma} - \Gamma^\sigma_{\sigma l}),$$

the equations

$$^*R_{(ik)} - \lambda g_{(ik)} = 0 \tag{10.10a}$$

$$^*(R_{[ik]} - \lambda g_{[ik]})_{,l} + (^*R_{[kl]} - \lambda g_{[kl]})_{,i} + (^*R_{[li]} - \lambda g_{[li]})_{,k} = 0 \tag{10.10b}$$

$$\mathfrak{g}^{[ik]}{}_{,k} = 0 \tag{10.10c}$$

represent the final field equations. $^*R_{kl}$ is formed from $^*\Gamma^m_{kl}$. Symmetric indices are in parentheses, antisymmetric ones in brackets.

For $\lambda = 0$ the equations reduce to the ones of Einstein and Straus. Although now the claim to include the nuclear forces was dropped, Schrödinger was satisfied by the closeness of his results to the ones of Einstein and Straus (1946).

Equation (10.10a) refers to the gravitational field, whereas (10.10b) and

(10.10c) describe the electromagnetic field and its interaction with the gravitational field. With $g^{[ik]} = 0$ (10.10a) yields the field equation of general relativity for the gravitational field with a cosmological term. In this case the $*\Gamma^i_{kl}$ become the usual Christoffel symbols.

13.11. The electromagnetic field in its final formulation

It was always one of the fundamental postulates of unified field theory that after switching off the other fields, i.e. after returning to a Riemannian metric, the equations of general relativity should remain and should describe the gravitational field. This requirement evidently is satisfied by (10.10a–c).

It is not possible, however, to satisfy a similar demand for the electromagnetic field, in the sense that after switching off gravitation in (10.10b) and (10.10c) the equations for a pure electromagnetic field should appear. Schrödinger investigated (10.10b) and (10.10c) in the proximity of a pseudo-Euclidean metric (Schrödinger, 1951) and found 'that a pure, chargefree Maxwellian field is capable of producing a gravitational field which according to the old theory could only be produced by matter other than an electromagnetic field'. He continued: 'This raises the hope, that in this theory we may be able to picture ordinary matter without sticking it in explicitly'.

On the other hand, there is also a reaction of the electromagnetic field to the gravitational field, and the point is made in Schrödinger (1947) that (10.10a–c) 'ought to explain the magnetic field produced by a rotating mass as the earth or the sun'. To end with, a paper by Schrödinger together with the present author shall be mentioned (Hittmair and Schrödinger, 1951), in which the propagation of light in strong electric and magnetic fields was investigated. By using Lorentz-covariance one succeeds in deriving an equation for the velocity of light which corresponds to the one in an anisotropic medium.

During this work, the author received a lasting impression of how persistently Schrödinger applied himself to a problem which promised conclusive consequences of his unified field theory

13.12. Conclusions

In the following years, solutions to Schrödinger's field equations were found by several authors, see, for example, Ikeda (1954, 1955). They refer to spherical symmetry as well as to plane wave form. No physically tangible results were obtained, however. Schrödinger's Annual Institute Report of 1954–5 puts it this way: 'It is a disconcerting situation that ten years endeavour of competent theorists has not yielded even a plausible glimpse of Coulomb's law.' If it could still be hoped then that exact solutions would lead the way to physical significance (Kilmister and Stephenson, 1954), this hope has vanished nowadays as quite a number of such solutions are known. One has to admit that classical unified field theory is not in the line of current physical thinking; it is 'geometrically overloaded but physically undernourished' (Goenner, 1984).

In the years following 1953, after ten years of intensive work, Schrödinger turned away from the subject of unified field theory. A close acquaintance of his, Bruno Bertotti, wrote: 'These years [1953–5] marked an important turning point in his scientific outlook and work: he abandoned his research on unified field theory and turned completely to his philosophical and epistemological interests.'

In those years, Schrödinger invested a lot of ingenuity and arduous work in a high goal. His writings excelled by preciseness and vivacity of expression. His book *Space-Time Structure* (Schrödinger, 1960), which emerged at that time, renders these qualities in a particularly impressive way.

The development of unified field theory, however, has taken a different direction. The old unified field theory has widely fallen into oblivion by now. The acronym GUT, however, reminds, although nowadays in a modified significance, of Schrödinger's heroic attempt to attain the great unity of physical laws by way of geometrizing classical fields. High expectations were followed by pronounced disillusionment in this period of a great scientist's life.

13.13. Acknowledgements

The author is particularly indebted to Dr. Auguste Dick who obligingly made accessible quotations from letters in Schrödinger's posthumous papers. Also the Central Library of the Physics Institutes in Vienna under the direction of Dr. W. Kerner was helpful in every way. Thanks are due to Profs. H. Urbankte and M. Schweda for valuable comments and for manuscripts, and to the colleagues of the Physics Institute (T. U. Vienna) who read the draft. In many regards the study of Schrödinger's writings was facilitated by the *Collected Papers*, edited by W. Thirring (Vieweg and Son, Vienna, 1984).

References

Born, M. and Infeld, L. (1934) *Proc. Roy. Soc. A* 144, 425–51
Eddington, A. S. (1923) *The Mathematical Theory of Relativity*. Cambridge University Press

Einstein, A. (1923) *Sitzungsber. Preuss. Akad. Wiss. (Berl.)*, pp. 32–8; 76–8; 137–42
Einstein, A. and Straus, E. G. (1946) *Ann. Math. Stud.* 47, 731
Goenner, H. F. (1984) *Proc. Sir Arthur Eddington Centenary Symp.* 1, 176–96
Hittmair, O. and Schrödinger, E. (1951) *Comm. Dublin Inst.* A8, 1–15
Ikeda, M. (1954) *Progr. Theor. Phys.* 12, 17–30
Ikeda, M. (1955) *Progr. Theor. Phys.* 13, 265–75
Kilmister, C. W. and Stephenson, G. (1954) *Suppl. Nuov. Chim.* 11, 118–40
Mie, G. (1912a) *Ann. Phys.* 37, 511–34
Mie, G. (1912b) *Ann. Phys.* 39, 1–40
Mie, G. (1913) *Ann. Phys.* 40, 1–66
Schrödinger, E. (1943a) *Proc. Roy. Irish Acad.* 49A, 43–58
Schrödinger, E. (1943b) *Proc. Roy. Irish Acad.* 49A, 135–48
Schrödinger, E. (1944) *Proc. Roy. Irish Acad.* 49A, 275–87
Schrödinger, E. (1946) *Proc. Roy. Irish Acad.* 51A, 41–50
Schrödinger, E. (1947) *Proc. Roy. Irish Acad.* 51A, 163–71
Schrödinger, E. (1951) *Comm. Dublin Inst.* A6, 1–28
Schrödinger, E. (1960) *Space-Time Structure.* Cambridge University Press
Weyl, H. (1918) *Raum, Zeit, Materie.* Berlin. (Translated as *Space, Time, Matter* by H. L. Brose (1922). Methuen, London)

14

The Schrödinger equation of the Universe

S. W. HAWKING

University of Cambridge

The Schrödinger equation is usually thought of as governing the behaviour of matter on a small scale. By a small system may be meant anything from two particles up to a whole star. Here, I want to consider a slightly larger system, the Universe. As has been remarked elsewhere, Schrödinger's equation comes into its own when classical physics breaks down. An example of breakdown on a small scale was provided by the classical model of the atom. Classical physics predicted that the electron would spiral into the nucleus and matter would collapse. Indeed, quantum mechanics and Schrödinger's equation were invented precisely to overcome this problem. There is a similar problem with the Universe. Classical physics predicts that there was a time about ten billion years ago when the density of matter would have been infinite. This is called the Big Bang singularity, and most people take it to be the beginning of the Universe. However, here I want to report some recent work which shows that, if one applies the Schrödinger equation to the whole Universe, there is no singularity. Instead one gets a wave function which corresponds in a classical limit to a Universe which starts from a minimum radius, expands in an inflationary manner at first, goes over to a matter dominated expansion, reaches a maximum radius and collapses again.

In the case of a single particle, the wave function Ψ is a function of position x and time t. It obeys the Schrödinger equation

$$\frac{\partial \Psi}{\partial t} = -iH\Psi$$

$$= -i\left[\frac{-\partial^2}{\partial x^2} + V(x)\right]\Psi(x, t).$$

I shall use units in which $\hbar = c = 1$.

One can extend this to the case of a scalar field ϕ_0. The wave function Ψ is now a functional of the scalar field configurations $\phi(x)$ on a surface of constant time t such that $|\Psi|^2$ is the probability of finding the given scalar field configuration on the 3-surface of constant time. It obeys a functional Schrödinger equation

$$\frac{\partial \Psi}{\partial t} = -iH\Psi$$

$$= -i \int \left[\frac{-\delta^2}{(\delta\phi(x))^2} + m^2\phi^2(x) \right] d^3x \, \Psi[\phi(x)](t).$$

In the case of gravity the wave function Ψ is a functional of the metric h_{ij} on a spacelike 3-surface and of the matter field configurations on that surface such that $|\Psi|^2$ is the probability of finding a 3-surface with the metric h_{ij} and the given matter field configuration. If the 3-surface is noncompact and asymptotically flat, the wave function Ψ is also a function of the asymptotic time coordinate t of the 3-surface at infinity. However, it seems that our Universe is not asymptotically flat. We do not know whether it is spatially finite or infinite, but I shall assume the former possibility. In that case, the wave function does not depend on time explicitly because there is no invariant definition of time in general relativity. Instead, the metric h_{ij} determines the position of the 3-surface in time because it determines where one can fit a 3-surface with the given metric into a 4-geometry. Because the wave function does not depend on time explicitly, the Schrödinger equation has the very simple form

$$0 = H\Psi$$

$$= \left[-G_{ijkl}\frac{\delta^2}{\delta h_{ij}\delta h_{kl}} - h^{1/2}\left({}^3R(h) + 2\Lambda + 8\pi m_p^{-2}T_{nn}\left(\frac{\delta}{\delta\phi}, \phi\right) \right) \right]\Psi[h_{ij}, \phi],$$

where G_{ijkl} is the metric on superspace, the space of all 3-metrics h_{ij}

$$G_{ijkl} = \tfrac{1}{2}h^{1/2}(h_{ik}h_{jl} + h_{il}h_{jk} - h_{ij}h_{kl}).$$

${}^3R(h)$ is the Ricci scalar for the 3-metric h_{ij} and T_{nn} is the normal component of the energy-momentum. This equation, which is also called the Wheeler–DeWitt equation, can truly be called the Schrödinger equation of the Universe. If we can solve it, we can determine the quantum state of the Universe and everything in it. The 'metric' G_{ijkl} of superspace is a product of matrices with signature $(-+++++)$ for each point of the 3-surface. By choosing a suitable gauge condition, one can arrange that G_{ijkl} has a signature which has one minus and an infinite number of pluses. Thus one can regard the Schrödinger equation of the Universe as an infinite dimensional hyperbolic equation on superspace. The role of time coordinates is played by the size of the 3-manifold. Thus, if one knows the form of the wave function Ψ for small 3-surfaces, one can solve the Schrödinger equation as a Cauchy problem to determine the wave function for larger 3-surfaces. Very small 3-surfaces correspond to very early times in the Universe near the Big Bang singularity that is predicted by classical general relativity. Thus the problem of the boundary conditions of the Schrödinger equation of the Universe is closely related to that of the initial conditions of the Universe.

In ordinary quantum mechanics, one can determine the ground state of the Schrödinger equation by a path integral prescription. For example, in the case of a simple harmonic oscillator with position x and time t the ground state wave function is given by a Euclidean path integral

$$\Psi(x') = \int d[l] \, e^{-\hat{I}[l]}$$

where the integral is over all paths $l(\tau)$ ($\tau = it$) such that $l(0) = x'$ and $l(-\infty) = 0$, and $\hat{I}[l]$ is the action of the path l in the Euclidean space with coordinates (x, τ). Similarly in field theory one can determine the ground state wave function $\Psi[\phi(k)]$ by a Euclidean path integral

$$\Psi[\phi'(x)] = \int d[\phi] \, e^{-\hat{I}[\phi]}$$

where the integral is over all field configurations $\phi(x, \tau)$ in Euclidean space such that $\phi(x, 0) = \phi'(x)$ and $\phi(x, -\infty) = 0$ and $\hat{I}[\phi]$ is the Euclidean action of the configuration $\phi(x, \tau)$.

Following this approach one can define the wave function of the Universe by an integral over positive definite metrics

$$\Psi[h_{ij}, \phi'] = \int_C d[g_{\mu\nu}] \, d[\phi] \, e^{-\hat{I}[g_{\mu\nu}, \phi]}$$

where the integral is over all positive definite 3-metrics $g_{\mu\nu}$ and matter field configurations ϕ which belong to some class C and which are bounded by a 3-surface on which the metric is h_{ij} and the matter field configuration is ϕ'. The specification of the class C determines the quantum state of the Universe, i.e. it picks out one particular solution of the Schrödinger equation.

The most natural choice of the class C of positive definite metrics in the path integral for the wave function of the Universe might seem to be metrics that are asymptotically flat at large distances, i.e. they would be asymptotic to the standard flat metric on Euclidean space. However, it turns out that this boundary condition does not provide a suitable wave function for the Universe. In fact, one finds that the path integral is dominated by disconnected 4-metrics which consist of a compact part joined on to the 3-surface and a separate asymptotically Euclidean metric. It therefore seems more natural to choose the class C to consist only of compact metrics. This incorporates the idea that the Universe is completely self-contained: there is no external asymptotic region. One can paraphrase this by saying that the boundary conditions of the Universe are that it has no boundary.

The real test of a proposed boundary condition for the Universe is whether it predicts a universe like the one that we observe. It is too difficult to solve the infinite dimensional Schrödinger equation of the Universe, but we can get some idea by freezing out all but a finite number of degrees of freedom. The Schrödinger equation then becomes a finite dimensional equation that can be solved with the given boundary conditions. The results in simple models are very encouraging; for example, in the case of a spatially closed universe of radius a containing a massive scalar field ϕ, the Schrödinger equation is

$$\left(\frac{\partial^2}{\partial a^2} - \frac{1}{a^2}\frac{\partial^2}{\partial \phi^2} - a^2 + a^4 m^2 \phi^2\right)\Psi(a, \phi) = 0.$$

The boundary conditions imply that Ψ is regular as $a \to 0$, the classical Big Bang singularity. This is like the case of an atom: the wave function is regular at the nucleus even though classical physics predicts a singularity. If one solves the Schrödinger equation with the boundary conditions one gets a wave function which can be interpreted as a superposition of quantum states peaked around classical solutions which represent inflationary universes and which could be a good model for the universe that we live in.

15

Overview of particle physics*

A. SALAM

International Centre for Theoretical Physics, Trieste

15.1 Introduction

Erwin Schrödinger and those in his scientific generation – men like Werner Heisenberg and P. A. M. Dirac – were interested in physics because it provides us with the basic understanding of the laws of nature. Thus, even though their work has provided us with the basis of most of modern high technology, particularly through the emergence of the quantum theory of the solid state, their pursuit of physics was not motivated by this. In this sense we, in our generation, look upon Erwin Schrödinger's situation and that of his colleagues with a degree of envy.

In keeping with the tradition of Erwin Schrödinger, I have pleasure in presenting this overview of particle physics as in early 1986. I am sure this overview will be outdated by the time it sees print – but that is the fate of anything one may write in so fast moving a subject.

Physics is an incredibly rich discipline: it not only provides us with the basic understnading of the laws of nature, it is also the basis of most of modern high technology. This remark is relevant to our developing countries. A fine example of this synthesis of a basic understanding of nature with high technology is provided by liquid-crystal physics which was worked out at Bangalore by Prof. S. Chandrasekhar and his group. In this context, one may note that, because of this connection with high technology and materials' exploitation, physics is the 'science of wealth creation' par excellence. This is even in contrast to chemistry and biology which – though as important for development – are 'survival sciences'. This is in the sense that chemistry in application is concerned with fertilizers, pesticides, etc., while biology is concerned with medical sciences. Thus,

* This overview was first given at the International Conference on Mathematical Physics, University of Chittagong, Bangladesh, and Second Asia Pacific Physics Conference, Bangalore, India, 1986.

together, chemistry and medical sciences provide the survival basis of food production as well as of pharmaceutical expertise. Physics takes over at the next level of sophistication. If a nation wants to become wealthy, it must acquire a high degree of expertise in physics, both pure and applied.

15.2. Overview of particle physics

In the past, particle physics was driven by a troika which consisted of (*a*) theory, (*b*) experiment, and (*c*) accelerator and detection-devices technology. To this troika have been added two more horses. Particle physics is now synonymous with (*d*) early cosmology (from 10^{-43} s up to the end of the first three minutes of the Universe's life) and it is strongly interacting with (*e*) pure mathematics. One may recall Res Jost who made the statement (towards the end of the 1950s) that all the mathematics which a particle physicist needed to know was a rudimentary knowledge of Latin and Greek alphabets so that one can populate one's equations with indices. No more now.

The situation in this regard has changed so drastically that now a theoretical particle physicist must know algebraic geometry, topology, Riemann surface theory, index theorems and the like. The more mathematics that one knows, the deeper the insights that one may aspire for.

In the last decade or so, in particle physics, we have been experiencing an age of great syntheses and of great vitality. At the same time, this is an age of great danger for the future of the subject in the sense that we need higher and higher accelerator energies, and more costly non-accelerator and passive experiments which take a higher injection of funds as well as of experimentation time, for discovering new phenomena or for testing the truth or the inadequacy of theoretical concepts. This is in contrast to the time when I started research (late 1940s and early 1950s) when we had ever-increasing quantities of undigested experimental data, but little *coherent* theoretical corpus of concepts.

15.3. Three types of ideas

I shall divide my remarks into three topics: (*a*) ideas which have been tested or will soon be tested with the accelerators which are presently being constructed; (*b*) theoretical ideas whose time has not yet come, so far as the availability of accelerators to test them goes; and (*c*) passive experiments which have tested – but not conclusively so far – some of the theories of the 1970s. To give a brief summary of what I want to dwell on, consider each of these three topics:

Ideas which have been tested or will soon be tested include the standard model based on the symmetry group $SU_c(3) \times SU_L(2) \times U(1)$, with which there is no discrepancy known at the present time; light Higgs which may be discovered at SLC after 1987 or at LEP after 1989; preons of which quarks may be made up. Because of the low momentum transfers involved, it is very unlikely, but preons may be discovered at HERA (after 1991) and may fetch a totally new slant to bear on the family problem, and on the problem of quark elementarity.

Theoretical ideas whose time has not yet come (basically because no accelerators are being constructed to test them) include $N = 1$ supersymmetry and $N = 1$ supergravity – the lower limit for supersymmetric partners for presently known particles appears to be rising and may now be as large as 50 GeV. Persuasive theoretical arguments would lead us to expect that such supersymmetric partners of quarks and leptons may exist below 1 TeV. To find these, we shall need LHC (large hadron collider in the LEP tunnel), or SSC (super conducting supercollider being considered in the USA), or an e^+e^- collider with centre of mass energy in the TeV range. For any of these, there has been no sanction from the European or US Governments. If at all, such accelerators may not arrive before the years 1995–2000.

Other related ideas in this category which also need higher energies are right-handed weak currents, extended supergravities, and super string ideas. I shall discuss these later.

The set of ideas for which non-accelerator and passive experiments have been mounted are mostly concerned with grand unification of electroweak and strong forces in its multifarious ramifications. These include proton decays, $n\bar{n}$ oscillations, $(\tau_{n\bar{n}} > 10^6 \text{ s } 90\% \text{ confidence limits})$, neutrino masses and oscillations, monopoles, dark matter and the like. A number of these experiments have been tried, but not with much success so far.

Let us now turn to each of these topics in turn.

15.3. Ideas which have been tested or will soon be tested

Since in this context we shall be concerned with the early availability of accelerators, I shall start with Table 15.1 which gives a list of already existing, soon to be commissioned, as well as the proposed accelerators.

While we are discussing the availability of future accelerators, let me make the following remark. The highest electric field gradients (which determine the size of an accelerator) achievable with today's technology, are no higher than 0.1 GV m^{-1}. Twenty years hence, when we may have perfected the technology of laser beat-wave plasma accelerators, this gradient may go up by a factor of 1000, i.e. $0.1\,\mathrm{TV\,m^{-1}}$. This may mean that a 30 km accelerator may produce $\sqrt{s} \simeq 10^4\,\mathrm{TeV}$. An accelerator

Table 15.1. *Accelerations in the foreseeable future*

Year	Machine	$\sqrt{s}$ (GeV)	Constituent $\sqrt{s}$ (peak-max, GeV)	Luminosity	Locality
1986	SppS	900	$100-200_{q\bar{q},q\bar{q}}$	$>10^{30}$	CERN
1986	Tevatron	2000	$200-600_{q\bar{q},q\bar{q}}$	$<10^{31}$	FERMILAB
1987	TRISTAN (e^+e^-)	60	60	$<10^{32}$	Japan
1987	SLC (e^+e^-)	100	100	10^{30}	Stanford
1987	Bepc (e^+e^-)	4	4		Beijing
1989	LEP (I) (e^+e^-)	100	100	10^{31}	CERN
?	LEP (II) (e^+e^-)	200	200	10^{31}	CERN
1990	UNK	3000	$300-900_{q\bar{q},q\bar{q}}$		Serpukhov
1991	HERA (eq)	320	$100-170$	$>10^{31}$	Hamburg
?	LHC	8000–16 000	2000		CERN
?	SSC	40 000	4000	10^{33}	USA
?	e^+e^-	4000	4000	10^{33}	?

Note the important role of luminosity in Figs. 15.1 and 15.2 due to G. L. Kane, which exhibit the windows for discovering heavy quarks or leptoquarks and which show that the construction of the SSC with its higher luminosity (as well as higher energy) is imperative.

Fig. 15.1. Windows for given mass (m), defined to detect the effect barely, not to study it. $\langle x \rangle = (1/5)\sum_{j,Q,W^+W^-,H,\tilde{g}}(m/\sqrt{s})$. Compare energy, luminosity and beams; could use $2m$, or geometric mean, so absolute size of $\langle x \rangle$ is somewhat arbitrary.

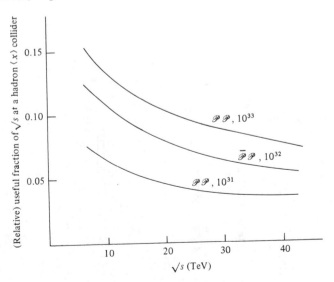

circling the moon may generate 10^6 TeV. (This was suggested by Arthur C. Clarke.)

An accelerator circling the Earth, as Fermi once conceived, may be capable of $\sqrt{s} \simeq 10^7$ TeV, while an accelerator extending from Earth to the Sun would be capable of $\sqrt{s} \simeq 10^{11}$ TeV. In the same crazy strain, for an accelerator to be capable of generating $\sqrt{s} \simeq 10^{16}$ TeV (the theoretically favoured, Planck mass m_P) one would need 10 light years! Clearly one must eventually fall back on the highest energy cosmic rays to study, for example, the likes of the recently discovered high energy muon signals in the Nusex (Mont Blanc) and Soudan I experiments (Fig. 15.3). These muons (produced in the atmosphere) can apparently be traced back to a cosmological accelerator associated with Cygnus X-3 – an X-ray source discovered in 1966, some 37 000 light years distant from us, which has a duty cycle of 4.8 h and an integrated luminosity of 10^5 Suns.

From the muon signals, these recent Nusex and Soudan experiments have claimed that Cygnus X-3 is beaming to us radiation of a new kind (light photinos, neutral light quark nuggets?) of energy 10^4 TeV. (I shall not discuss here why most of the familiar particles are ruled out.)

Fig. 15.2. Windows for heavy quarks, squarks or leptoquarks.

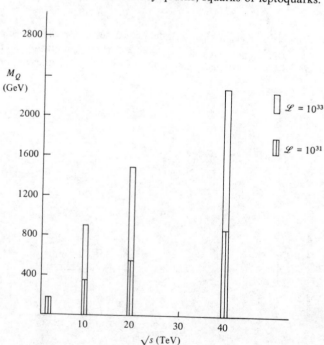

If this experimental evidence is taken at its face value, how is the radiation beamed at us by Cygnus X-3 generated? One speculative idea is that the Cygnus system may consist of a binary – a conventional main sequence star and a pulsar or a black hole. Matter from the conventional star accretes around the compact pulsar or the black hole, forming a disc. The protons thus accelerated (up to maximum energies of $\simeq 10^5$ TeV) go into a beam dump, where is created the mysterious radiation, which hits our atmosphere and makes the observed muons.

One interesting aspect of the situation is that (as warned by Halzen at the Bari Conference in summer, 1985), the swan may be dying (Fig. 15.4) – the

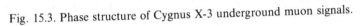

Fig. 15.3. Phase structure of Cygnus X-3 underground muon signals.

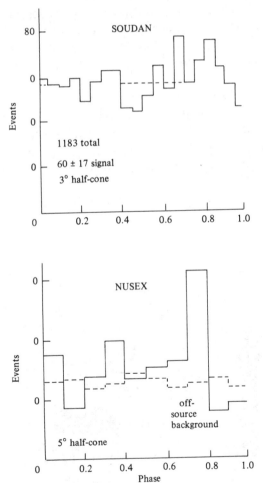

emitted flux seems to be decreasing at the rate of a decade over three years, 'much as if the beam dump was being blown away'.

Are there likely to be more intense, more energetic sources than Cygnus X-3 in the sky? Will cosmology come to the rescue of experimental particle physics?

15.5. The standard model and the light Higgs

The standard model of today's particle physics describes three families of quarks and leptons. The first family consists of (u_L, d_L) and (u_R, d_R) quarks; each quark comes in three colours: red, yellow and blue. There are, in addition, three leptons, (e_L, ν_L) and e_R. Thus this family has 12 quarks and three leptons (i.e. 15 two-component objects).

The second family has charm and strange quarks (replacing the up and down quarks) while the electron and its neutrino are replaced by the muon and its neutrino. Like the first family, there 15 two-component objects. The third family likewise consists of top and bottom quarks plus the tauon and its neutrino.

In addition to these 45 spin $\frac{1}{2}$ two-component objects there are the 12 Yang–Mills–Shaw gauge spin 1 mediators corresponding to the symmetry $SU_c(3) \times SU_L(2) \times U(1)$ – the photon γ, $W^\pm$, Z^0 and eight gluons. Of these, nine particles (γ and eight gluons) are massless. In addition, there should be one physical spin zero Higgs H^0 giving a total of 118 degrees of freedom for

Fig. 15.4. Time-dependence of the photon flux from Cygnus X-3.

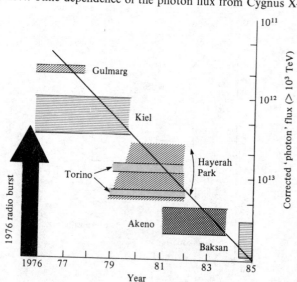

the particles in the standard model. All particles except the Higgs in this list have been discovered and their masses and spins determined (though the top-quark is still disputed). In this context it is worth remarking that CERN data on Sp$\bar{\text{p}}$S, has confirmed the theoretical tree-level expectation of $W^{\pm}$, Z^0 masses to within 1%.

So far as the Higgs particle is concerned, theory does not give its mass. Defining a *light Higgs* as an object with a mass $< 300 \text{ GeV}$ and a heavy Higgs as an object with a mass beyond, up to 1 TeV, one may remark that a heavy Higgs would have a large width. Thus the concept of a particle for it would be lost. (In case the Higgs is heavy, the W and Z would interact strongly. One would expect a new spectroscopy of bound states and Regge trajectories, which may include spin 1 resonances and which would modify the properties of W and Z. No one likes this possibility, but it could happen.)

Fig. 15.5 shows the possible signals of the standard model Higgs. As one can see, beyond a mass of 60 GeV, one would need LEP II and eventually the SSC supercollider.

One of the measurements which has been carried out during 1985, relevant to the number of families in the standard model, is the estimate of the number of *light* neutrinos which may couple to the Z^0 particle. This

Fig. 15.5. The possible signals of the standard model Higgs. Beyond a mass of 60 GeV one would need LEP II and eventually the SSC supercollider. (G. Kane.)

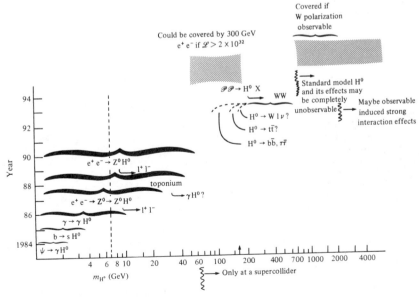

number has been estimated from the collider measurements (on Z^0 width) to be $<5.4\pm$ a few – amazingly consistent with the number 3 or 4 which cosmological data would appear to favour. This cosmological data is deduced from He^3 and He^4 abundances. Re-examining this data (and taking into account its errors) Ellis *et al.* have suggested that cosmology may even be consistent with five or six light neutrinos (three light v_R in addition to v_L?). No longer can one say with Landau 'cosmologists are seldom right, but never in doubt'. They could be right this time!

An important set of experiments which would be carried out at SLC and at the LEP accelerator concern the radiative corrections to the tree-level predictions of the standard model in the electroweak sector. As an example, Table 15.2.

Lynn, B., Peskin, M. and Stuart, R. (1985), SLAC-PUB-3725

One loop physics	$\delta A_{LR}=\delta A_{\tau pol}$	δA_{FB}	δM_W (MeV)
QED vacuum polarization	-0.13	-0.075	-800
GSW weak with $m_t=30$ $M_H=100$	-0.03	-0.01	-180
Heavy top quark $m_t \simeq 180\,GeV$	0.03	0.0075	780
Heavy Higgs $\simeq 1$ TeV	-0.01	-0.0045	-160
Heavy quark pair (a) large I splitting	0.02	0.01	300
(b) degenerate	-0.004	-0.002	-42
Heavy lepton pair (a) large I splitting	0.012	0.006	300
(b) degenerate	-0.0013	<0.001	-14
Heavy squark pair (a) large I splitting	0.02	0.01	300
(b) degenerate	0	0	0
Heavy Slepton pair (a) large I splitting	0.012	0.006	300
(b) degenerate	0	0	0
Winos (a) $m_{3/2}<100\,GeV$	0.005	0.0025	100
(b) $m_{3/2}>100$ GeV	<0.001	<0.001	<10
Technicolor $SU_8 \times SU_8$	-0.04	-0.018	-500
O_{16}	-0.07	-0.032	-500
Strong interaction uncertainty	±0.0030	±0.0016	±25

in Table 15.2 are presented data due to Lynn, Peskin, Stuart and Verzegnassi relevant to these radiative corrections.

The important asymmetries to be measured to high precision (possibly $< 1\%$) on Z^0 resonance at SLC/LEP are the initial electron beam polarization asymmetry A_{LR} and the forward-backward asymmetry A_{FB} in $e^+e^- \to \mu^+\mu^-$ and the final state τ^- polarization asymmetry $A_{\tau pol}$ in $e^+e^- \to \tau^+\tau^-$. Another precision measurement to be done at LEP II or the Tevatron collider is the precise $W^{\pm}$ mass. These shifts in these quantities $\delta A_{LR}, \delta A_{\tau pol}, \delta A_{FB}$ and δM_W (listed in Table 15.2) are the difference between the tree plus one-loop calculations including the relevant physics listed in column 1 of the table and the quantity calculated at tree level taking α, G_{μ} (muon decay constant) and M_Z (precise Z^0 mass) as known precisely measured input parameters. M_Z will be measured to high precision ($\sim \pm 50$ MeV) at LEP/SLC as will of course the precise Z^0 width M_Z which changes by ~ 180 MeV for each light neutrino species and can, therefore, be used to 'count' light neutrinos as indicated above. The comparison of theory and experiment at the loop level would, in analogy with the Lamb shift in the late 1950s, vindicate the renormalizability of non-Abelian gauge theories as a principle of nature.

Assuming that Z^0 mass will be measured with extreme accuracy at SLC or LEP (up to 50 MeV or possibly better), the parameters of the standard model could be chosen as (α, G_F, m_Z) and $(m_t$ and $m_H)$. One can thus propose clean tests of the electroweak theory at the one loop level. These could consist of measurements of longitudinal polarization, measurement of W mass (LEP II), and measurement of neutrino $\sigma(\nu e)/\sigma(\bar{\nu}e)$ ratios (charm 2) to one loop level.

Consider the case of the longitudinal polarization in

$$A_{LR} = \frac{\sigma_{e^+e_L^-} - \sigma_{e^+e_R^-}}{\sigma_{e^+e_L^-} + \sigma_{e^+e_R^-}} \to \mu\bar{\mu}.$$

On top of the Z^0 resonance, the full one loop prediction is $A_{LR} = -0.03$ for $m_H = 100$ GeV. A (new) heavy quark pair would contribute $+0.02$, a heavy scalar lepton pair another $+0.012$ and so on. Thus one may hope to determine from the comparative measurements of $\delta A_{LR}, \delta M_W$ etc., the top quark mass or the Higgs mass or the existence of new heavy quark pairs etc. in an indirect fashion.

15.6. Ideas whose time has not yet come

$N=1$ supersymmetry and $N=1$ supergravity

The most important ideas of this category are of $N=1$ supersymmetry and $N=1$ supergravity. $N=1$ supersymmetry is the hypothetical symmetry (between fermion and bosons) which decrees that a spin 1/2 must be accompanied by two spin zeros: a spin 1 gauge particle must be accompanied by a massless spin 1/2 particle (gaugino): a massless spin 2 graviton must be accompanied by a massless spin 3/2 gravitino, and so forth. (For $N=2$ extended supersymmetry, one would group in one multiplet, two spin 0, two spin 1/2's and one spin 1 object. For the maximal $N=8$ extended supersymmetry, there is just one multiplet containing one spin 2, accompanied by eight spin 3/2, 28 spin 1, 56 spin 1/2 and 70 spin 0 states.)

Supersymmetry is an incredibly beautiful theory – a compelling theory if there is one, even though there is no physical evidence of the existence of supersymmetry partners to the known particles, up to 30 GeV (or even perhaps up to 50 GeV).

One aspect of its compellingness lies in its superior renormalizability properties and the possibility which these open up of understanding why the large numbers which occur in particle physics could be 'naturally' stable.

Consider as an example the large number $m_P/m_W \simeq 10^{17}$, where m_P is Planck mass. (Planck mass occurs in gravity theories: large numbers similar to m_P/m_W can however occur in all grand unification theories which synthesize electroweak with strong forces.) Only in supersymmetric theories can one show that such a number, once fixed at the tree level, would be unaffected by radiative corrections. This is one of the virtues of supersymmetric theories.

But supersymmetry must be a highly broken symmetry. What is the supersymmetry breaking mass? Or more physically, where do the missing supersymmetry partners of quarks, leptons, photons, $W^\pm$ and Z^0 lie? The theoretical expectation seems to be *below* a (flexible) upper limit of 1 TeV, if supersymmetry is relevant to the electroweak phenomena. (To one loop for example, $\delta m_H^2 = \alpha/\pi |m_F^2 - m_B^2|$; if $\delta m_H^2 \simeq m_W^2$, we get the estimate that $|m_F - m_B| \leqslant 1$ TeV.)

What is the mass limit at present of supersymmetry partners not having been found? Estimates vary, but for a conservative recent (model-dependent) estimate, see Fig. 15.6. Here m_0 and $m_{1/2}$ are two supersymmetric parameters in terms of which $m_{\tilde{q}}$, $m_{\tilde{l}}$, $m_{\tilde{W}}$, $m_{\tilde{Z}}$, etc. can be parameterized. (The symbol $\sim$ denotes a supersymmetric partner.) For example $m_{\tilde{q}}^2 \simeq m_0^2 + 7m_{1/2}^2$, $m_{\tilde{l}}^2 \simeq m_0^2 + 0.5m_{1/2}^2$, $m_{\tilde{g}} \simeq 3m_{1/2} \simeq 7m_{\tilde{l}}$. Ellis has examined all relevant data and concluded that possibly the lower limit for supersymmetric partners may be as high as 50 GeV.

To conclude, it is expected that supersymmetry may make itself manifest with highly luminous accelerators with centre of mass energy in excess of 1 TeV (e.g. LHC, SSC or an e^+e^- linear collider of >1 TeV). Supersymmetry may manifest itself at lower energies as an indirect phenomena, (supersymmetry was claimed with monojets, dijets, trijets at UA1, but the present backgrounds happen to be too large to draw unambiguous conclusions).

Supersymmetry and $N=1$ supergravity
Note the following points:

(1) The $N=1$ supersymmetrization of the standard model will need two multiplets of Higgs particles (plus, of course, their Higginos).
(2) The signature of supersymmetry is the R quantum number which is defined as $+1$ for all known particles and -1 for their supersymmetric partners. Thus these new partners must be produced in pairs, and among the expected supersymmetry particles there must be a lowest mass stable object which must be neutral in order to survive the Big Bang. Further, it must be weakly coupled otherwise it will be concentrated in condensed form in the galaxies. The favourite candidates for this object are scalar neutrinos $\tilde{\nu}$, photinos $\tilde{\gamma}$ or gravitinos – the spin 3/2 partners of the gravitons.
(3) If $N=1$ supersymmetry comes, $N=1$ supergravity cannot be far behind.

Fig. 15.6. Compilation of experimental constraints of SUSY breaking parameters. (Presented by J. Ellis at Kyoto Conference, September 1985.)

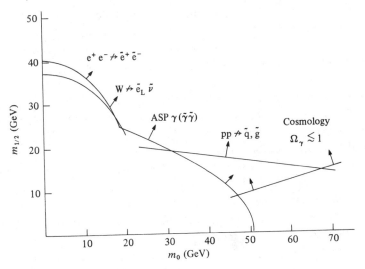

The argument goes as follows: the major theoretical question regarding supersymmetry is supersymmetry breaking. The only recent known way to break supersymmetry is to break it spontaneously. For this to work, one starts with a gauge theory of supersymmetry, i.e. a supergravity theory which (for the $N = 1$ case), would contain spin $3/2$ gravitinos in addition to spin 2 gravitons. One would then postulate a super-Higgs effect, i.e. a spin $1/2$ and spin zero matter multiplet (of 'shadow' matter which interacts with known particles only gravitationally). The spin $1/2$ member of this multiplet would be swallowed by the spin $3/2$ gravitino – the latter becoming massive in the classic Higgs fashion to break supersymmetry spontaneously. The (mass)^2 of the gravitino (in analogy with the standard Higgs effect) would be of the order of the gravity coupling parameter $(1/m_P^2)$ times the expectation value of the supersymmetry breaking potential $(m_{3/2}^2 \simeq 1/m_P^2 (0|V|0))$.

(4) What could be the numerical estimates for the supersymmetry breaking parameter $(0|V|0)$? Chamseddine et al., Weinberg, and Nilles et al. and others noted (1982–3) that with the reasonable requirement of a vanishing cosmological constant, one finds that this (spontaneously broken $N = 1$) supergravitization of the (supersymmetry) standard model automatically leads to a breaking of $SU_L(2) \times U(1)$. Thus, $m_{3/2} \simeq m_W$ is motivated, and with this one may estimate, $(0|V|0)^{1/4} \simeq 10^{10}$ GeV. (An alternative no scale model which automatically generates a zero cosmological constant is the favourite at CERN. Here $m_{3/2}$ is not tied to m_W.)

Unification of gravity with other forces

So far we considered $(N = 1)$ supergravity as following on the heels of $(N = 1)$ supersymmetry in order to provide for an orderly breaking of supersymmetry – there was no true unification of gravity with other forces. Let us now discuss a true unification of gravity with the rest of particle physics.

The first physicist to conceive of this and to try to find experimental evidence of such a phenomenon was Michael Faraday. In a symbolic drawing (Fig. 15.7), one may see the equipment which Faraday used at the Royal Institution in Piccadilly, London (and which is on exhibition even today). The failure of his attempts did not dismay Faraday. He wrote afterwards: 'If the hope should prove well founded, how great and mighty and sublime in its hitherto unchangeable character is the force I am trying to deal with, and how large may be the new domain of knowledge that may be opened to the mind of man.'

The first semi-successful theoretical attempt (in the 1920s) to unify gravity with electromagnetism was that of Kaluza and Klein who showed in a theory based on a five-dimensional space-time, that the appropriate curvature component in fifth dimension, corresponds to electromagnetism.

Fig. 15.7. Cartoon due to A. de Rujula.

Further, if the fifth dimension is compactified to a scale R, and charged matter is introduced, one can show that α and G must be related as $\alpha \simeq G/R^2$. These ideas were beautifully generalized in an extended supergravity context, when Cremmer and Julia discovered in 1979 that the extended $N = 8$ supergravity in four dimensions emerges as the zero mass limit of the compactified $N = 1$ supergravity in 11 dimensions. An incredible achievement. Since 1979, all supergravitors have lived in higher dimensions.

At that time, this theory was hailed as the first TOE (Theory of Everything). The compactification of $d = 11$ Kaluza–Klein supergravity down to four dimensions would give as its zero mass sector, gravitons as well as gauge particles like spin 1 γ, $W^\pm$ and Z as well as the 56 fermions in one multiplet of $N = 8$ supergravity. Unfortunately, the $N = 8$ theory and this particular multiplet suffered from two fatal defects: the fermions were not chiral and the theory did not have the content of the standard model so far as quarks, leptons or even $W^\pm$ were concerned. In addition to the zero mass sector, there would, of course, be higher Planck mass particles $(\text{mass})^2 \simeq$ multiples of $1/R^2$ – the so-called pyrgons, providing an embarrassment of riches.

Can one ever obtain direct evidence for the existence of higher dimensions? The answer is, possibly, yes. If the extra dimensions happen to be compactified today, why should they remain so? Why should they not share the universal expansion? Could $\dot{R} \neq 0$ since α, G and R are expected to be related to each other? If we are fortunate and if $\dot{\alpha}/\alpha$ and/or $\dot{G}/G$ should turn out to be non-zero at the present experimental level, this might most simply be explained by postulating extra dimensions. At present, the experimental limits are less than 10^{17} years for $\dot{\alpha}/\alpha$, while $\dot{G}/G$ is less than 10^{11} year^{-1}.

Anomaly-free supergravities

Where do we stand today so far as higher dimensions and extended supergravity theories are concerned? It would appear that the only theories which may combine chiral fermions and gravity are $N = 1$ in $d = 10$ dimensions or $N = 2$ in $d = 6$ or 10. In order that such theories contain the known chiral quarks and leptons as well as the W's and Z and photons and gluons the most promising is the $N = 1$, $d = 10$ supergravity, but it would have to be supplemented with a supersymmetric Yang–Mills multiplet of matter in addition to the supergravity multiplet. Likewise, the $N = 2$ theory in $d = 6$ would need, not only an extra Yang–Mills field, but also nonlinearly realized (σ-model) matter fields. Thus a pure Kaluza–Klein supergravity will never be enough. Higher dimensions, yes; but to generate the known gauge theories of electroweak and strong forces, we need (higher dimensional) super-Yang–Mills in addition.

As if this was not trouble enough, both $d=6$ or $d=10$ theories were shown to be anomalous and gravitationally replete with infinities. This impasse was broken only in Autumn, 1984, by Green and Schwarz who showed that $d=10$ supergravity with Yang–Mills $SO(32)$ or $E_8 \times E_8$ could be made anomaly-free by an addition of certain numbers of new terms. They further showed that these additional terms were already present in the spinning supersymmetric string theories in ten dimensions, invented earlier. (Whether these string theories were free of gravity infinities beyond one loop was, and still is, an open question.) And this brings us to the new world of superstrings and the new version of a TOE.

Spinning supersymmetric strings

Consistent, Lorentz-covariant and conformally symmetric string theories had been written down in 2, 10 or 26 dimensions (the last being relevant only to bosonic strings) already in the 1970s. A closed string is a loop which replaces a space-time point. Its quantum oscillations correspond to particles of higher spins and higher masses, which may be strung on a linear trajectory in a Regge spin versus mass^2 plot. If the slope parameter of this trajectory – the only parameter in the theory – is adjusted to equal Newtonian constant, one can show, quite miraculously, that in the zeroth order of the closed bosonic string, there emerges, from the string theory, Einstein's gravity in its fullness! (The higher orders give $(\text{Planck})^{-1}$ range corrections.)

Further, the supersymmetric ten-dimensional spinning string theory (descended from 26 dimensions) could exist in a 'heterotic' form, invented by Gross *et al.*, with a built-in Yang–Mills gauge symmetry with a gauge group G of rank 16 which can be either $G = SO(32)/Z_2$ or $E_8 \times E_8$. This theory, though chiral, is anomaly-free. The descent from 26 to ten dimensions is accomplished by compactification on a 16 torus $(26-10=16)$, which, using the beautiful results of Frenkel and Kac, in fact reproduces the full tally of 496 Yang–Mills massless gauge particles associated with $SO(32)/Z_2$ or $E_8 \times E_8$ even though we started with only 16 gauge particles corresponding to the 16 tori. The remaining 480 gauge particles start life as solitons in the theory. The theory is gravitational (and chiral) anomaly-free. The hope is that this theory may also be finite to all loop orders: the only finite theory of physics containing quantum gravity.

Can we proceed from ten down to four physical dimensions? Witten *et al.* have shown that the ten-dimensional theory can indeed be compactified to four-dimensional Minkowski space-time × an internal six-dimensional space with $SU(3)$ holonomy (a Calabi–Yau space) which preserves a residual $N = 1$ supersymmetry in four dimensions. A number of families emerge; their count is equal to 1/2 of the Euler number of the compactified space. The Yukawa couplings allowed in the theory are expected to be topologically determined.

It is all these remarkable features of superstring theories which make the string theorist 'purr' with deserved pride.

String theory as the TOE

Could this be the long-awaited unified theory of all low energy phenomena in nature? There is a fair prospect of this. But would such a theory be a TOE? The answer in my opinion is no. As remarked before, all theories which descend from higher to lower dimensions must contain massive particles in multiples of Planck mass $m_P \simeq 1/R \simeq (\alpha/G)^{1/2}$. Since no *direct* tests of existence or interactions of such objects can ever be feasible, there will always remain the experimentally unexplored area of these higher masses and energies – in addition, of course, to the mystery of the quantum of action. What I am saying is that before they can be called a TOE, one must prove, at the least, a uniqueness of our present theories – a theorem which states that if a theory fits all known phenomena at low energies, it can have only *one* extrapolation to higher energies. From all past experience, this is unlikely, even as regards to the framework. (Think of the framework of Newtonian gravity versus that of Einstein's gravity.)

There arises an important question at this stage whether the string theories represent a wholly new attitude so far as the fundamental theory is concerned, or whether we are dealing with a relativistic quantum field theory of the familiar type in two dimensions.

The latter point of view has been argued for by a number of authors: by Polyakov and those who have followed him, and latterly by Weinberg and in Trieste. This is the point of view of the so-called 'first-quantized' string, represented by two-dimensional gravity theory where the conventional Einstein action is the Euler number. The 'string' Lagrangian of Nambu, Gotto, Nielson, Susskind and others (which for the bosonic case is represented by a set of 26 fields), is the preonic-matter Lagrangian, in interaction with this two-dimensional (non-propagating) gravity. Physics arises through topology which is determined by the number of handles on a two-dimensional sphere, on which the theory lives. (The number of these handles is specified by the Einstein action, which, as I said earlier, is the Euler number for the manifold in two and only two dimensions.) A phase transition which makes the 26 preonic fields acquire expectation values predicates the transition to the 'familiar' chart of $d = 26$, space and time. Like all such theories, this relativistic quantum field theory is expected to be unitary so far as the basic 26 preonic fields are concerned. (Conformal invariance plays a crucial role in this.)

This formalism must be supplemented by composite-field expressions (for example, for the likes of $W^\pm$, Z^0, etc.), in terms of the basic 26 preonic fields (vertex operators which must respect conformality). Expressions for these can be written down in terms of products of preonic fields and their derivatives. The unitarity of this composite field theory presents a familiar challenge – familiar in the sense that this problem is on par with the unitarity problem of theories of composite hadrons, protons and neutrons – in the context of a fundamental quark theory.

The other point of view favours the 'second quantized' version of the string theory. Here, unitarity problems for the likes of W's, Z's, photons (and also quarks and leptons in a spinning string version), present no difficulties. (It is as if one was writing the local field theory of hadrons.) Within the light-cone framework (the only one where the full supersymmetric version of the second quantized theory exists) the fact that 'local' field theories of extended objects, like strings, should exist at all, is highly non-trivial. That such theories (of open and closed strings) number no more than seven altogether, is an added bonus.

Perhaps the profoundest aspects of this second quantized formulation are represented by Witten, who derives the basic interaction in such a theory (at least for the purely bosonic open string), within a non-commutative geometry context – freed of dependence on space-time charts. If this point of view succeeds, the string theories would be ushering in an era, like that of quantum theory in 1925 and 1926 when a new epistemology came into existence to replace the humbler point of view represented by the 'old' quantum theory.

Which of these two points of view will yield deeper insights, and in the end prevail? Time will tell.

But apart from these matters of interpretation, the one crucial question which our experimental colleagues are entitled to ask, is this: what are the compelling experimental consequences of string theories?

The emergence of (necessarily a supersymmetric) standard model with the right number of families may, of course, be a triumph, but will it establish the superiority of the string attitude? At present, there are few unambiguous *new* predictions. One of them concerns the existence of one or two new Z^0's. In Fig. 15.8 is shown the window for such Z^0's.

Unfortunately, their masses, even their existence, are not firmly predicted by the theory. A possibly firmer and more spectacular prediction is the existence of fractionally charged non-confined dyons which would, of course carry the appropriate integral magnetic monopolarity in accordance with the Dirac formula.

15.7. Passive and non-accelerator experiments

Next we come to the passive experiments which are mainly concerned with testing cosmological gauge aspects of grand unification theories (unifying electroweak and strong nuclear interactions). These are tests for (i) monopoles (topological defects of the π_2 type). Though predicted by the gauge theories concerned at high temperatures prevailing in the early Universe, one would not like too many monopoles around now (note the claimed detection during 1985 of the South Kensington monopole): (ii)

Fig. 15.8. Window for new Z^0 (absolute scale model dependent).

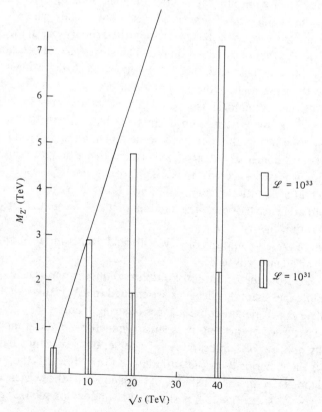

cosmological strings (π_1) which are good for galaxy seeding; and (iii) domain walls (π_0), which apparently are a cosmological disaster. Surely, this set of predictions presents a mixed bag of desirables and undesirables! In addition there is the question mark on varieties of remnant dark matter, endemic to a vast variety of theories and whose ever-lengthening list is given in Table 15.3. (I shall not dwell on the role of inflation in cosmology in this context, which apparently can also resolve the problem of over-abundance of monopoles.)

Among the most celebrated passive experiments is proton decay. As we know, a limit on $P \to e^+ \to \pi^0 > 2.5 \times 10^{32}$ years partial decay-time is suggested by the IMB collaboration. There are, however, claims for (seven) candidate-events for $P \to e^+ + K^0$ and $N \to v + \eta^0$ and $N \to v^0 + K^0$ modes, by the Kiolar Gold Fields collaboration, Kamiokande and Nusex. (A firm detection of K's would signal supersymmetry and also explain the longer life-time.) A worrisome background is due to atmospheric neutrinos, which would make it difficult, on Earth, to be sure of a real signal for proton decay if its life much exceeds 10^{33} years. Pati, Sreekantan and Salam have suggested experiments on the Moon where even though the primary flux of cosmic rays is unhindered by the existence of an atmosphere or magnetic fields, an experiment carried out in a tunnel or a cavern with 100 m of moon-rock surrounding it on all sides, would cut down the backgrounds (in particular of v_e neutrinos) to a figure less than 1/100 of the background on Earth. If the proton life-time lies within the range of 10^{33} and 10^{35} years,

Table 15.3. *Expected dark matter*

	Mass	Origin (time, temperature)
Invisible axion	10^{-5} ev	10^{-30} s (10^{12} GeV)
v	30 ev	1 s (1 MeV)
Light $\tilde{\gamma}$, gravitino	~kev	10^{-4} s (100 MeV)
Heavy $\tilde{\gamma}$ gravitino,		
Axiano, sneutrino, v	~GeV	
Monopole	10^{16} GeV	10^{-34} s (10^{14} GeV)
Kaluza–Klein particles and shadow matter (maximons, pyrgons, etc.)	10^{18}–10^{19} GeV	10^{-43} s (10^{15} GeV)
Quark nuggets	10^{15} g	10^{-5} s (300 MeV)
Primordial black holes	$> 10^{15}$ g	$> 10^{12}$ s

experiments on the Moon may become necessary for unambiguous detection.

The cost of such moon experiments consists in taking around 50 tons of detecting material to the moon, plus the cost of the making of the cavern; it may come to around $1 billion. Such outlays would become feasible if moon colonization programmes are pursued seriously. I have no doubt that this will happen if there is a banning of nuclear weapons, since technological, advanced societies must spend funds on high technology projects, in order to keep the overall economy healthy.

This concludes my brief overview of particle physics. The progress in this field would surely have pleased Erwin Schrödinger.

16

Gauge fields, topological defects and cosmology

T. W. B. KIBBLE

Imperial College, London

The types of topological defects that can appear in gauge theories – domain walls, strings and monopoles – are described in the following. The possibility that such defects were generated at phase transitions in the very early history of the Universe is discussed, in particular the idea that cosmic strings may provide the seeds for the density perturbations from which galaxies form.

16.1. Introduction

The subject that I want to discuss did not exist in Erwin Schrödinger's time. But it forms a natural bridge between two of his major interests – the fundamental physics of the smallest and the largest scales.

As Yang describes in this volume (pp. 53–64) Schrödinger contributed very significantly to the early development of gauge theories – the fundamental framework for all our present understanding of elementary particles and their interaction. He also made influential contributions to cosmology.

At the time these two fields were essentially separate. Indeed one might say that our failure so far to reconcile the basic physical theories of the small and the large is the major outstanding problem in physics. Schrödinger was clearly aware of the lack. He was very interested in any attempt to bridge the gap, for example in Eddington's ambitious though unsuccessful fundamental theory (Schrödinger, 1938).

A final reconciliation of relativity and quantum theory still eludes us, but cosmology and particle physics are no longer poles apart. Even 20 years ago they were still largely unrelated disciplines, each unfamiliar to the practitioners of the other, save for a few bold spirits like Schrödinger. Now they are closely interwoven. No student of particle physics or cosmology can afford to be wholly ignorant of the other field.

I want to discuss one particular bridge between the two, an idea that came from studies of gauge theories and may well be important in understanding the large-scale structures of the Universe. This is the notion of a topological defect.

I shall begin by trying to describe what these defects are and how they arise, and then I shall go on to explain their possible relevance to cosmology, concluding with a discussion of one specific and very exciting idea, namely that the so-called cosmic strings may be responsible for the observed lumpiness of our Universe – the fact that matter is distributed in galaxies and clusters rather than being spread more uniformly.

16.2. Topological defects

There are several kinds of defects, including domain walls, strings or vortices, monopoles, textures and various composites such as walls bounded by strings and strings terminated by monopoles. The most interesting, however, seem to be strings, and I shall concentrate on these.

Consider the simplest example of a gauge theory – a theory of a complex scalar field Φ interacting with a gauge field described by the potential A_μ. The Lagrangian density is

$$L = D_\mu \Phi^* D^\mu \Phi - \tfrac{1}{4} F_{\mu\nu} F^{\mu\nu} - V(\Phi),$$

where

$$F_{\mu\nu} = \partial_\mu A_\nu - \partial_\nu A_\mu$$

and the covariant derivative is

$$D_\mu \Phi = \partial_\mu \Phi + ig A_\mu \Phi;$$

g is the gauge coupling or charge. Note that I use units in which $c = \hbar = 1$.

For the potential V we take

$$V = \tfrac{1}{2} h^2 (\Phi^* \Phi - \tfrac{1}{2} \eta^2)^2 \tag{2.1}$$

(where h and η are constants), which is invariant under the phase rotations

$$\Phi \to \Phi \, e^{i\alpha}. \tag{2.2}$$

This theory exhibits a phase transition at a critical temperature T_c roughly of order η, from a high-temperature symmetric phase to a low-temperature phase in which the symmetry (2.2) is spontaneously broken. Below T_c, Φ acquires a vacuum expectation value; in tree approximation at $T = 0$,

$$\langle \Phi \rangle = \frac{1}{2^{1/2}} \eta \, e^{i\alpha}. \tag{2.3}$$

The magnitude of $\langle \Phi \rangle$ is fixed, but there is a degenerate set of equivalent vacua distinguished by the phase angle α.

Consider a configuration in which as one traverses a loop in space the angle varies smoothly from 0 to 2π. Continuity of $\langle \Phi \rangle$ then demands that at some interior point $\langle \Phi \rangle$ vanishes. In fact there must be a line along which the expectation value is zero. This is a string.

In the context of relativistic field theory the string solution was first written down by Nielsen and Olesen (1973). A static string along the z axis is described by $\langle \Phi \rangle = \Phi(\rho, \phi)$, with

$$\Phi(\rho, \phi) = \frac{1}{2^{1/2}} \eta f(\rho) \, e^{i\phi}. \tag{2.4}$$

Hence $f(0) = 0$ and $f(\infty) = 1$. To ensure that $D_\mu \Phi = 0$ at large distances there must also be a gauge field, with an azimuthal component A_ϕ. One then finds a quantized 'magnetic' flux along the string

$$\int_0^\infty 2\pi \rho B_z(\rho) \, d\rho = \frac{2\pi}{g}, \tag{2.5}$$

where $B_z = -F_{xy}$.

The string has a tension μ equal to its energy per unit length. (This equality is a consequence of invariance under Lorentz boosts along the direction of the string.) The string tension μ is of order η^2, multiplied by a numerical constant depending on the ratio of the coupling constants.

Similar solutions are well known in many-body physics – in particular the quantized flux tube in a type-II superconductor and the vortex line in a superfluid, carrying quantized vorticity.

The solution is topologically stable. It cannot be deformed into one of the vacuum solutions (2.3) by any change in a finite region of space; the quantized flux (2.5) is conserved.

There do exist similar solutions with ϕ in (2.4) replaced by $n\phi$, and carrying flux $2\pi n/g$, for all integers n. In general, however, the solutions with $|n| > 1$ need not be stable; an $n = 2$ string may divide into two $n = 1$ strings.

16.3. Strings in non-Abelian gauge theories

In the example above the gauge group was the Abelian group $U(1)$, but strings can also appear in the spontaneous breaking of a non-Abelian symmetry. Consider a scalar field Φ belong to some representation σ of a gauge group G. If one possible vacuum state is described by

$$\langle \Phi \rangle = \Phi_0, \tag{3.1}$$

then other equivalent vacua correspond to other points on the orbit through Φ_0,

$$M = \{\sigma(g)\Phi_0 : g \in G\}.$$

If H is the isotropy subgroup defined by Φ_0,

$$H = \{h \in G : \sigma(h)\Phi_0 = \Phi_0\},$$

i.e. the subgroup of unbroken symmetries in the vacuum state given by (3.1), then the points on the orbit may be identified with left cosets of H in G:

$$M = \frac{G}{H}.$$

In the example, $G = U(1)$ and H consists of the identity element alone, so that M is a circle, S^1. The reason that stable strings exist in that case is that there are loops in M that cannot be shrunk to a point without leaving it. In fact, such loops are labelled by an integer n, the winding number, the same integer that determines the quantized flux. Stated more mathematically the first homotopy group of M is the additive group of integers

$$\pi_1(M) = Z.$$

If G is connected and simply connected, i.e. if $\pi_0(G)$ and $\pi_1(G)$ are both trivial, then according to a well-known theorem

$$\pi_1\left(\frac{G}{H}\right) = \pi_0(H) = \frac{H}{H_c}, \tag{3.2}$$

where H_c is the connected component of H. In other words, strings can exist if and only if the subgroup H of unbroken symmetries contains disconnected pieces. (To apply this criterion to the Abelian case, one must replace $U(1)$ by its simply connected covering group, $G = R$, the additive reals, in which case $H = Z$.)

An example in a model that has been considered as a realistic candidate for a unified gauge theory is the breaking of the $SO(10)$ group to $SU(5)$ by a Higgs field Φ in the 126-dimensional representation. Here we must take the two-fold covering group,

$$G = Spin(10).$$

We then find (Kibble, 1982) $H = SU(5) \times Z_2$, where Z_2 is the additive group of integers modulo 2, generated by a rotation through 2π. The existence of this discrete subgroup implies that

$$\pi_1(M) = \pi_0(H) = Z_2.$$

Hence there are strings in the model, but in this case a string has no fixed orientation; a string can be deformed through configurations of finite energy per unit length into an antistring.

16.4. Other defects

The other classes of defects are also characterized by appropriate homotopy groups (Kibble, 1976).

Domain walls occurs when a discrete symmetry is broken, so that M has disconnected pieces; equivalently, $\pi_0(M)$ is non-trivial. The simplest example is a *real* field Φ with a double-well potential similar to (2.1). Physically, the domain wall is a thin slice separating regions where Φ is in one or other of the disjoint minima. Away from the wall Φ is close to a minimum, but as one crosses the wall it varies rapidly.

Next suppose that $\pi_2(M)$ is non-trivial. This means there are closed surfaces in M that cannot be shrunk to a point within M, for instance when M is a two-sphere. In this case we may have a trapped point defect or monopole; if we have a closed surface such that the corresponding values of Φ trace out an unshrinkable surface in M, then somewhere *within* the spatial surface must be a region where Φ leaves M – the monopole.

If G is simply connected then as for strings the classifying group may be identified with a homotopy group of H, namely

$$\pi_2(M) = \pi_1(H). \tag{4.1}$$

This is non-trivial in particular if H contains a $U(1)$ factor. Since we know that the exact unbroken symmetry group that is manifest at low energies does contain such a factor, associated with electromagnetic gauge transformations, we should expect monopoles to appear at *some* transition.

Finally, there is an interesting phenomenon that can occur when $\pi_3(M)$ is non-trivial. This is the appearance of what are called textures – topologically distinct configurations with no singularity or localized concentration of energy associated with them. In a closed Universe the possible states would fall into disjoint classes labelled by the elements of $\pi_3(M)$. However, it is not at all clear that there would be any observable consequences (but see Davis, 1986).

16.5. The early Universe

Two of the most striking pieces of observational evidence about the large-scale structure of the Universe are the Hubble expansion and the cosmic microwave background. Distant galaxies are receding from us with velocity

$$v = Hr.$$

Observationally, the Hubble parameter H lies in the range 50–100 km s^{-1} Mpc^{-1}. Clearly at an earlier epoch the Universe was much denser than it now is.

Measurements of the microwave background have revealed a nearly isotropic blackbody spectrum, with a temperature of about 2.7 K. (There is a dipole anisotropy, interpreted as due to our peculiar velocity of about 600 km s^{-1} relative to the cosmic rest-frame, but when this is subtracted, the remaining distribution appears uniform (Mandolesi *et al.*, 1986; Uson and Wilkinson, 1984).) This radiation is interpreted as the red-shifted relic of blackbody radiation emitted when the Universe was denser and hotter, in fact at about 3500 K, the temperature above which atomic hydrogen would have been ionized.

If we follow the evolution backwards to yet earlier times, the Universe must have been even denser and hotter. Over much of its early history, it would have contained a plasma of highly relativistic particles, expanding and cooling adiabatically. However, it is likely that this smooth expansion was punctuated by a sequence of phase transitions. The last two of these are fairly well understood. When the temperature was about 100 GeV (in units where Boltzmann's constant is unity), the Universe went through the electroweak transition, in which the $SU(2) \times U(1)$ symmetry of Glashow, Salam and Weinberg broke to the $U(1)$ of electromagnetism. At about 100 MeV there followed the quark–hadron transition at which a dense soup of quarks and gluons broke up into a gas of hadrons.

16.6. Defects in the early Universe

From the present viewpoint, however, these are not the interesting transitions. For the appearance of topological structures we must look to earlier, more speculative transitions associated with the breaking of larger, unified symmetries, or perhaps of supersymmetry.

To be specific, let us consider a transition at which a $U(1)$ symmetry is broken as a complex scalar field Φ acquires a non-zero expectation value. In most regions of the Universe, once the temperature has fallen well below the critical temperature T_c, the magnitude of Φ will be close to its vacuum value. However, as in (2.2), its phase is arbitrary and will take on randomly different values in different regions. Occasionally it will happen that around some loop the phase changes by 2π. Then somewhere within the loop must be a trapped defect, a string, along which Φ is prevented from reaching its vacuum value but instead has to vanish. The result will be to generate a random tangle of strings (Kibble, 1976). Since the strings in this model cannot terminate, they either form closed loops or are infinite in length.

In just the same way, in other models random configurations of domain walls or monopoles would be produced. Both these latter possibilities are disastrous from a cosmological point of view. Domain walls if present would rapidly come to dominate the energy density of the Universe, yielding an extremely inhomogeneous Universe expanding at an accelerating rate, quite unlike what is actually observed (Zel'dovich *et al.*, 1974).

Monopoles once produced can disappear only by annihilation with anti-monopoles. This process cannot be sufficiently rapid to reduce their number density to acceptable levels by the present (Preskill, 1979). Unless the monopoles were very light and formed at a relatively recent transition, they too would come to dominate the Universe and make it expand too fast.

If monopoles or domain walls were produced at a high-temperature phase transition they must somehow have been eliminated – perhaps by the process of 'inflation' in which the Universe undergoes a period of exponential expansion. The inflationary Universe model was proposed to solve a number of cosmological problems, among them the 'monopole problem' (see, for example, Linde, 1984). The exponential expansion so dilutes the number density of monopoles (or anything else present before inflation) as to make it unobservably small.

Strings, as I shall explain, do not necessarily pose the same kind of problem. If inflation is invoked to solve the monopole and domain wall problems, it would also eliminate strings. But it is consistent to suppose that strings were formed at another transition after inflation.

16.7. Evolution of cosmic strings

Let us assume that there is a phase transition at which strings appear, at a temperature somewhere between the electroweak scale (100 GeV) and the Planck scale, at which gravity becomes strong,

$$m_P = G^{-1/2} \simeq 10^{19} \text{ GeV},$$

where G is Newton's constant. A convenient dimensionless parameter characterizing the gravitational interactions of the string is

$$G\mu \simeq \left(\frac{T_c}{m_P}\right)^2. \tag{7.1}$$

The range being considered is

$$10^{-34} < G\mu < 1,$$

but as we shall see the most interesting range from a cosmological point of view is around 10^{-6}.

At the phase transition a random tangle of strings is produced. A typical string will have a 'Brownian' configuration characterized by a persistence length ξ related to the correlation length for the phase of Φ. Initially the strings move in a dense medium and are heavily damped (Kibble, 1976). Small kinks in the string will tend to straighten, thus increasing ξ and decreasing the total length of string.

The subsequent evolution depends critically on what happens when strings cross. Two things might happen; the strings might simply pass through one another or they might exchange partners, creating two new kinks, which will straighten out in turn. Calculations that have been done suggest that they do normally exchange partners except when their relative velocity at crossing is very large, in excess of perhaps $0.9c$.

If the probability of exchanging partners is large, the characteristic length scale ξ will grow rapidly and the total length of string will decrease. This continues until ξ is of the same order as the so-called horizon distance or the age t of the Universe. This happens (Everett, 1981; Kibble, 1982) when the temperature has fallen to

$$T_* \simeq G\mu m_P \qquad (7.2)$$

(cf. $T_c = (G\mu)^{1/2} m_P$).

From this point onwards the strings are no longer heavily damped and can acquire relativistic speeds. Evolution of the string configuration must then lead to one or other of two possible final states – string domination or a scaling solution.

If the strings as a whole do not lose much energy they will eventually come to dominate. A string-dominated Universe is not as extreme as one dominated by domain walls but is in fact quite similar to one dominated by non-relativistic matter. It is just within the bounds of possibility that our Universe is string-dominated, but only if the strings are very light $(G\mu < 10^{-12})$ and came to dominate only rather recently, in particular after the time of nucleosynthesis of the light elements (Kibble, 1986; Vilenkin, 1984).

The second possibility, of a scaling solution, is a much more interesting one. What it means is that the configuration of strings would look essentially the same at any time (once scaling is established) but scaled in proportion to the horizon size. The characteristic length scale ξ will be a fixed fraction of t.

16.8. The role of loops

A scaling solution will be attained only if there is a sufficiently effective mechanism for transferring energy from the system of strings to other components of the Universe. Such a mechanism was first proposed by Vilenkin (1981b), namely the formation of closed loops and their subsequent decay by gravitational radiation. Subsequently it has been found that these loops may play a vital role in cosmology.

When a string intersects itself it can create a loop which is cut off and isolated from the rest of the string. Such a loop will oscillate with a period equal to one-half of its total length. As it oscillates it emits gravitational radiation. The radiated power turns out to be independent of the size of loop, equal to a small multiple of $G\mu^2$ (Burden, 1985; Turok, 1984; Vachaspati and Vilenkin, 1985). Consequently the loop will gradually shrink. If its initial size is l it will disappear entirely after a time of order $l/10G\mu$.

The gravitational radiation allows us to place an upper limit on $G\mu$. The emitted radiation would disturb the regular periodicity of the millisecond pulsar. Thus observations of its regularity allow us to place an upper limit (Hogan and Rees, 1984),

$$G\mu \lesssim 10^{-5},$$

a limit which should improve with time as the duration of the observations grows.

Since $G\mu \ll 1$, the lifetime of a loop is long compared to its size. It follows that at any given time there will be a large population of small loops surviving from a much earlier epoch. In fact the total energy density in the form of small loops is much larger than that in long strings. It is natural to ask whether these small loops might provide the seeds of the density perturbations from which galaxies eventually form.

The origin of these density perturbations is one of the major unsolved puzzles in cosmology. None of the existing theories is wholly satisfactory. This problem is closely related to another – the question of what constitutes the so-called dark matter in the Universe, matter whose dynamical influence we can detect but which we cannot identify with any visible objects. Broadly speaking there are two possibilities – hot dark matter and cold dark matter – though the names are rather misleading; they refer not to the temperature now but to whether or not the particles involved were effectively relativistic at the decoupling time (Primack, 1984).

The cold dark matter scenario can account reasonably well for the origin of galaxies but has difficulty explaining the observed structure on very large scales (voids and superclusters). On the other hand, with hot dark matter it is hard to see how galaxies could have formed in the time available, because the free streaming of relativistic particles tends to erase small-scale perturbations. What is needed to make the hot dark matter theory viable is some class of objects that do not get erased but survive to form the seeds for galaxy formation. Loops of string do the job admirably.

16.9. Tests of the comsic string hypothesis

The idea that galaxies form around loops can be tested. If true, the spatial distribution of galaxies should reflect the earlier spatial distribution of loops. This can be predicted from numerical simulations of the process of loop formation.

In addition to the small loops that seed the condensation of galaxies there will be rarer large loops, which may be expected to seed larger condensations. It is reasonable to suppose that the observed rare rich clusters of galaxies form around these large loops. This idea has been tested in work on the spatial correlation function $\xi(r)$ of rich clusters by Turok (1985). (The definition is as follows: given a cluster at the origin, the probability of finding another at distance r is enhanced by the factor $[1 + \xi(r)]$.)

The process of loop formation is essentially scale-independent. Hence, the correlation function $\xi_l(r)$ for loops of a given minimum size l can be written as a function of the dimensionless ratio r/d_l, where d_l is the mean separation of loops of size l:

$$\xi_l(r) = \xi\left(\frac{r}{d_l}\right).$$

The form of this function can be extracted from numerical simulations performed by Albrecht and Turok (1985). Over a considerable range, from 0.1 to about 4 it can be well approximated by

$$\xi\left(\frac{r}{d}\right) \simeq 0.2\left(\frac{r}{d}\right)^{-2}.$$

This may be compared directly with the observed correlation function of rich clusters, provided we set d equal to the mean separation of such clusters, namely 55 Mpc. The fit is astonishingly good. This is a very considerable achievement, because no other adequate explanation for this correlation function is known.

The theory allows an estimation of the parameter $G\mu$ from the requirement that loops of this mean separation be heavy enough to accumulate around themselves condensations of the size of rich clusters. Turok's estimate was

$$G\mu \simeq 2 \times 10^{-6}. \tag{9.1}$$

More recent work suggests that the value should be somewhat larger, perhaps 5×10^{-6}.

Similar tests can be applied to the correlation function of galaxies, and although there are some complicating features the results are in good agreement.

A number of other observational tests of the cosmic string scenario are available. Perhaps the most interesting is their gravitational lensing effect. A straight static string does not in fact exert any gravitational attraction on matter in its vicinity (Vilenkin, 1981a). This is an odd feature due to the fact that tension acts as a negative source of gravitational field, and in this case the tension just cancels the effect of mass or energy density. However, strings do have a gravitational effect. The space around a string is cone-shaped, as though a wedge of angle

$$\delta = 8\pi G\mu$$

had been removed and the faces glued together. Consequently one may view the same object twice around opposite sides of the string, and see two images separated by an angle which is typically $\delta/2$. Such pairs of images of quasars are known and may perhaps be due to strings (Hogan and Narayan, 1984; Kaiser and Stebbins, 1984; Silk and Vilenkin, 1984).

There are more prosaic explanations of this gravitational lensing, but an effect which if seen would be unambiguous evidence for strings is the appearance of a sharp discontinuity in the temperature of the microwave background radiation. If a string is moving across our field of view with transverse velocity v then (Kaiser and Stebbins, 1984) the apparent temperature on opposite sides will differ by

$$\Delta T \simeq 8\pi G\mu v T.$$

With a typical velocity of 0.4c and using Turok's value (9.1) for $G\mu$, this amounts to

$$\Delta T \simeq 50 \, \mu K,$$

but it could well exceed 100 μK. This is at most a factor of 3 below current limits.

16.10. Conclusions

The cosmic string theory has achieved some notable successes, though it has its problems too. Whether or not it turns out to be correct, it seems very likely that topological defects created at an early phase transition may have an important role to play in cosmology. This is another bridge between the physics of the very large and the very small, two fields that since Schrödinger's day have been coming ever closer.

References

Albrecht, A. and Turok, N. (1985) *Phys. Rev. Lett.* 54, 1868–71
Burden, C. J. (1985) *Phys. Lett.* 164B, 277–81
Davis, R. L. (1986) SLAC preprint SLAC-PUB-3896

Everett, A. E. (1981) *Phys. Rev.* D24, 858–68

Hogan, C. J. and Narayan, R. (1984) *Mon. Not. Roy. Astr. Soc.* 211, 575–91

Hogan, C. J. and Rees, M. J. (1984) *Nature* 311, 109–14

Kaiser, N. and Stebbins, A. (1984) *Nature* 310, 391–3

Kibble, T. W. B. (1976) *J. Phys. A* 9, 1387–97

Kibble, T. W. B. (1982) *Acta. Phys. Polon.* B13, 723–46

Kibble, T. W. B. (1986) *Phys. Rev.* D33, 328–32

Kibble, T. W. B., Lazarides, G. and Shafi, Q. (1979) *Phys. Lett.* 113B, 237–9

Linde, A. D. (1984) *Rep. Prog. Phys.* 47, 925–86

Mandolesi, N., Calzdari, P., Cortglioui, S., Delpino, F., Sironi, G., Inzani, P., De Amici, G., Solheim, J-E., Berger, L., Partridge, R. B., Martenis, P. L., Sangree, C. H. and Harvey, R. C. (1986) *Nature* 319, 751–3

Nielsen, H. B. and Olesen, P. (1973) *Nucl. Phys.* B61, 45–61

Preskill, J. P. (1979) *Phys. Rev. Lett.* 43, 1365–8

Primack, J. R. (1984) Lectures at the International School of Physics 'Enrico Fermi', Varenna, Italy, June 26–July 6, 1984

Schrödinger, E. (1938) *Il Nuovo Cimento* 15, 246–54

Silk, J. and Vilenkin, A. (1984) *Phys. Rev. Lett.* 53, 1700–3

Turok, N. (1984) *Nuc. Phys.* B242, 520–41

Turok, N. (1985) *Phys. Rev. Lett.* 55, 1801–4

Usòn, J. M. and Wilkinson, D. T. (1984) *Nature* 312, 427–9

Vachaspati, T. and Vilenkin, A. (1985) *Phys. Rev.* D31, 3052–8

Vilenkin, A. (1981a) *Phys. Rev.* D23, 852–7

Vilenkin, A. (1981b) *Phys. Rev. Lett.* 46, 1169–72, 1496 (erratum)

Vilenkin, A. (1984) *Phys. Rev. Lett.* 53, 1016–18

Zel'dovich, Ya. B., Kobzarev, I. Yu. and Okun, L. (1974) *Sov. Phys.-JETP* 40, 1–5; (*Zh. Exsp. Teor. Fiz.* 67, 3–11)

17

Quantum theory and astronomy

M.J. SEATON

University College, London

The physical conditions encountered in astronomy are often very extreme, and astronomy is therefore demanding in its use of known physical laws and constantly provides a stimulus for the discovery of new laws. Some of the conditions are also very simple.

The motion of a planet is simple compared with motions of terrestrial objects, and it was the study of this 'simple' problem which enabled Newton to formulate his laws of dynamics and gravitation and in so doing to found the subject of theoretical physics. Astronomy has continued to provide a main stimulus to studies of dynamics and of the nature of space, time and gravitation. Those, however, are topics for others. My brief is to discuss quantum theory and astronomy, and I will give some examples of advances in astronomy which have depended essentially on the use of quantum theory.

17.1. Stellar structure

Extreme conditions encountered in astronomy arise from high energies and temperatures, and densities which may be either very high or very low. I will first consider some problems of stellar structure. If $M(r)$ is the mass of a star interior to radius r, the law of mass conservation is

$$M(r) = \int_0^r \rho(r_1) 4\pi r_1^2 \, \mathrm{d}r_1 \tag{1.1}$$

where $\rho(r)$ is the density. Let $L(r)$ be the luminosity (total energy outflow) at radius r. For a static star the law of energy conservation is

$$L(r) = \int_0^r \varepsilon(r_1) 4\pi r_1^2 \, \mathrm{d}r_1 \tag{1.2}$$

where $\varepsilon(r)$ is the rate of energy production per unit volume, and the law of hydrostatic equilibrium is

$$\frac{dP(r)}{dr} = -\rho(r)g(r) \tag{1.3}$$

where $P(r)$ is the pressure and $g(r) = M(r)G/r^2$, the local gravitational acceleration. Derivation of a fourth equation, for the flux of radiant energy, is slightly more complicated. The equation of radiative transfer is

$$\frac{dI_\nu(r, \hat{s})}{ds} = -\kappa_\nu I_\nu + j_\nu$$

where $I_\nu(r, \hat{s})$ is the intensity of radiant energy at radius r in a direction $\hat{s}$, ds is a differential of distance in that direction, κ_ν is an extinction coefficient (the opacity), and j_ν is an emission coefficient. Conditions inside a star are not very different from those in a black-body enclosure, and one may take j_ν and κ_ν to satisfy Kirchhoff's law

$$\frac{j_\nu}{\kappa_\nu} = B_\nu(T) \equiv \frac{2h\nu^3}{c^2} \left[\exp(h\nu/kT) - 1\right]^{-1},$$

$B_\nu(T)$ being the intensity of black-body radiaton. Inside a star I_ν is nearly isotropic but not completely so because there is a net outward flux of radiation, F_ν, per unit area. Assuming near-isotropy the transfer equation can be solved to give

$$F_\nu = -\frac{4\pi}{3\kappa_\nu} \frac{dB_\nu}{dr}.$$

Integrating over ν gives the total flux F and the fourth equation of stellar structure,

$$F = -\frac{4\pi}{3\kappa} \frac{dB}{dr} \tag{1.4}$$

where

$$B = \int B_\nu \, d\nu = \frac{\sigma}{\pi} T^4$$

(σ being the Stefan–Boltzmann constant) and κ is the Rosseland mean opacity defined by

$$\frac{1}{\kappa} = \frac{\int (1/\kappa_\nu)(dB_\nu/dT) \, d\nu}{(dB/dT)}.$$

If all energy transport is by radiation, $L(r) = 4\pi r^2 F(r)$.

The four equations of stellar structure can be solved numerically, but it is instructive to obtain order-of-magnitude solutions using the crudest possible approximations. Using the known mass and radius of the Sun we deduce that its mean density is $\bar{\rho} = 1.4 \, \text{g cm}^{-3}$. The crudest approximation to the solution of the hydrostatic equation is

$$\frac{P_c}{R} = \bar{\rho}\bar{g}$$

where P_c is the central pressure, R is the total radius and $\bar{g}$ is a mean value of $g(r)$. This can be used to estimate P_c, and an estimate of the central temperature, T_c, is obtained using the perfect gas law,

$$P = NkT,$$

where N is the particle density and $\rho = N\mu$, where μ is the mean molecular weight. For the Sun this gives $T_c \simeq 2 \times 10^7$ K.

17.2. Energy-generation rate and opacity

The equations of stellar structure involve two quantities from basic physics, the energy-generation rate ε and the opacity κ. The determination of these two quantities requires a knowledge of quantum theory. If it is assumed that gravitational contraction is the only energy source for the Sun, calculations show that the time for which the Sun could maintain its present luminosity is much smaller than the known age of the Earth. The energy source must therefore be nuclear. In order to estimate values of ε which result from nuclear processes, a first task is to estimate the distance of closest approach for collisions of two nuclei in a stellar interior. Using our estimate T_c we obtain

$$kT_c \simeq \frac{\mu MG}{R}$$

(this also follows from the virial theorem: the mean kinetic energy equals the mean gravitational energy). For a head-on collision between two nuclei with charges Z_1, Z_2, the distance of closest approach, r_0, is such that

$$kT_c \simeq \frac{e^2 Z_1 Z_2}{r_0}.$$

It follows that

$$\frac{e^2 Z_1 Z_2}{r_0} \simeq \frac{\mu MG}{R},$$

this is to say that the mean potential energy at the distance of closest approach is equal to the mean gravitational energy. For collisions between two protons in the Sun this gives a value of r_0 which is much larger than the nuclear radius. The question which then arises is 'how can a nuclear reaction occur if there is no inter-penetration of the nuclei?' The answer, of course, is that it cannot but that quantum theory gives a finite probability of penetration into the classically forbidden region of $r < r_0$. This probability of barrier penetration is calculated from solutions of the Schrödinger equation for the two nuclei with a repulsive Coulomb potential. Once this is done one can begin to believe that the nuclear reactions do indeed provide the energy source for the Sun and the stars. A detailed theory of nuclear reactions in stars was first developed by Bethe and von Weizsäcker in 1938. For stars somewhat hotter than the Sun the main energy source is conversion of H to He in a catalytic chain of reactions involving C, N and O which starts with

$$^1H + {}^{12}C \rightarrow {}^{13}N + \gamma$$

involving the strong interactions. For the Sun, and stars cooler than the Sun, the barrier penetration probability for such reactions is still too small and the main energy source is conversion of H to He in a chain which starts with

$$^1H + {}^1H \rightarrow {}^2D + e^+ + \nu$$

involving the weak interactions. Many of the nuclear reactions in stars go *via* compound nuclear states (resonances), an example being

$$^1H + {}^{12}C \rightarrow ({}^{13}N)^{**} \rightarrow {}^{13}N + \gamma.$$

The energies involved are very low from the standpoint of experimental nuclear physics. The expressions used by astronomers for ε are usually obtained using phenomological theories of nuclear reactions together with experimental data on nuclear energy levels and on nuclear reactions to somewhat higher energies.

An order-of-magnitude estimate for the opacity κ in a stellar interior can be obtained from the approximate solution

$$F \simeq \frac{4\sigma}{3\kappa} \frac{T_c^4}{R}$$

of (1.4). From this we deduce that $\kappa \sim 10^3$ cm^{-1} for the Sun. Eddington had a gift for expressing such results in a homely way with a significance which can be easily appreciated. He converted such an estimate of κ to a value at the density of the Earth's atmosphere and, taking its reciprocal, obtained a photon path length of order 1 m which is really very small (a photon at optical wavelengths can traverse the entire atmosphere of the Earth with only a small probability for absorption). The opacity per unit mass is large for stellar interiors because one is concerned with ultra-violet photons in an ionized plasma. Many processes of absorption and scattering contribute to κ. The contribution from electron scattering is $\kappa(\text{electrons}) = N_e \sigma_T$, where N_e is the electron density and σ_T is the Thomson scattering cross-section. The contribution from free–free transitions

$$X^+ + e + h\nu \rightarrow X^+ + e$$

(thermal bremsstrahlung) can be calculated to a good first approximation using classical theory. Photo-ionization and spectral lines can also be very important even although only a small fraction of the nuclei have any bound electrons attached to them. Expressions for rates of radiative processes in hydrogenic ions were obtained by Kramers using the old quantum theory, and it was only two years after Schrödinger had first solved the wave equation for hydrogen that Gaunt obtained very general accurate expressions. For hydrogenic ions astronomers usually employ the expressions of Kramers multiplied by a Gaunt correction factor g. Despite an enormous amount of work having been done, a number of people, including myself, believe that more work is needed in calculating opacities, particularly the contributions from non-hydrogenic ions.

17.3. White dwarfs

A good example of extreme conditions in astronomy is provided by the white dwarfs (although they are not so extreme as the neutron stars). The designation 'white' implies that the spectral distribution of radiation is not very different from that of the Sun and hence that the surface temperature T_0 does not differ much from the solar value of $T_0 \simeq 6000$ K. Some white dwarfs have known distances and their outstanding characteristic is that they are very faint. Knowing the distance and the flux at the Earth one can estimate the luminosity L, and from the equation $L = 4\pi R^2 \sigma T_0^4$ deduce the radius. One finds that the white dwarf radius is about one-hundredth of the solar radius. Also some white dwarfs are found in binary star systems, and estimates of their masses can be obtained. The mass of a white dwarf is found to be comparable with the mass of the Sun, and it follows that the mean density must be larger by a factor of 10^6, giving $\bar{\rho}$(white dwarf) $\simeq 10^6$ g cm^{-3}. The great problem posed by the white dwarfs was that the pressure required to provide hydrostatic equilibrium is much larger than any value which can be reconciled with the perfect gas law. The solution to this problem was provided by the exclusion principle. At the density of a white dwarf all of the lower electron states are occupied and, even at low temperatures, electrons are forced to occupy states of high momentum, which gives high pressures. In 1983 a Nobel Prize was shared between W. A. Fowler for his work on nuclear reactions in stars and S. Chandrasekhar for his development, many years earlier, of the theory of white dwarfs.

17.4. Stellar spectra

Stellar spectra were first observed in the 1860s, and it caused great excitement to find that the stars contain the same chemical elements as are found on the Earth. Detailed analyses of astronomical spectra provide most of our knowledge of the classical composition of astronomical objects and much other information as well. If one starts from scratch the analysis must proceed in several states. First there is a 'coarse analysis' in which one attempts to get a correct general picture, what our American friends call 'the right ball park'. Only after this is done can one proceed to the 'fine analysis' which may involve a lot of detailed modelling. One result which emerges from the analysis, but is not apparent from casual inspection of stellar spectra, is that hydrogen is overwhelmingly the most abundant cosmic element. Helium comes next and all other elements have abundances which are much smaller.

Most lines in stellar spectra are in absorption, which is because the temperature in a star decreases outwards. All of the observed photons are produced in a star's surface layers, the 'stellar atmosphere'. For a photon with a frequency v in the vicinity of the frequency v_0 of a spectrum line, the opacity κ_v is large, the photon mean-free-path in the atmosphere is small and if the photon escapes it must have been produced in a high cool layer. For a coarse analysis it is assumed that the material particles in the open system of a stellar atmosphere obey the thermodynamic laws, at a local temperature, which are valid for a black-body enclosed system (this is known as local thermodynamic equilibrium or LTE). Using the thermodynamic laws we can calculate level populations, as functions of temperature and density, and if we also know the probabilities for radiative processes we can calculate the coefficients j_v and κ_v which, for LTE, satisfy Kirchhoff's law.

Stellar spectra were classified empirically, mainly by the Harvard astronomers, long before a satisfactory theory had been developed. Revisions were made as the work progressed which led to more jumbling of the letters of the alphabet. The accepted classification of spectra types is O, B, A, F, G, K, M. By considering only the colours of the stars this can be recognized as a temperature sequence, with the O stars being the hottest and most blue. The observed hydrogen lines are due to transitions between excited states and are of maximum strength for the A stars. The Boltzmann equation is

$$\frac{N_i}{N_1} = \frac{\omega_i}{\omega_1} \exp[-(E_i - E_1)/kT]$$

where N_i is the number density in the level with energy E_i, $i = 1$ is the ground level, and the ω's are statistical weights. This equation explains the weakness of the hydrogen lines in the cooler stars, which have temperatures such that the populations of excited states of hydrogen are very small. The hydrogen lines again become weak in the hottest stars because the hydrogen becomes ionized. The Boltzmann equation is not quantum mechanized in the sense that it does not involve Planck's constant and can be derived from classical statistics, but is in the sense that it involves the concepts of discrete energy levels and statistical weights. Derivation of the Saha equation of ionization equilibrium

$$\frac{N_i}{N_e N^+} = \frac{\omega_i}{2\omega_1^+}\left(\frac{h^2}{2\pi mkT}\right)^{3/2} \exp[I_i/kT],$$

(where N_e is the electron density, N_1^+ is the number density in the ground state of the next ionization stage, and I_i is the ionization energy for level i) involves a counting of quantum states of free electrons, and Planck's constant appears explicitly. It was this equation which enabled the interpretation of stellar spectral classifications to be completed, at least in broad outline.

There is something deceptively simple about the Boltzmann equation. Using it one finds that $\sum_i N_i$ is divergent, essentially because Coulomb potentials give an infinite number of bound states converging to the ionization limit. This is the problem of the divergence of the partition function and it is overcome on recognizing that at finite densities atoms cannot exist in arbitrarily high excited states (the mean atomic radii become larger than the mean separation between particles). These problems are of importance to astronomers who observe the disappearance of the hydrogen lines involving large principal quantum numbers.

One must consider the continuous spectra of stars as well as the line spectra. There is not too much difficulty for the hotter stars for which κ_ν has large contributions from electron scattering and from photo-ionization of ground and excited states of H, He and He^+, but there was a major problem for the cooler stars finally resolved by the suggestion of Wildt in 1939 that the photo-detachment process

$$H^- + h\nu \rightarrow H + e$$

is of major importance. A great deal of effort has gone into making accurate quantum mechanical calculations of the cross-section for this process. The best calculated results are probably correct to within 1 or 2%. Results are also available from laboratory measurements but their accuracy is probably not better than about 20%.

More detailed classification of stellar spectra is at least two-dimensional. In addition to the temperature classification $(O, B, A, \ldots)$ one must introduce luminosity classes depending on the radii of the stars, the giants and the dwarfs. The differences between spectra for stars belonging to different luminosity classes are mainly due to differences in pressure and hence in the widths of the spectrum lines. For the hot stars, having atmospheres which are partially ionized, the widths of the hydrogen lines are determined by the Stark effect, the electric fields being due to the charged particles in the atmospheres. Hydrogenic ions present a special case since they have a linear Stark effect resulting from the l-degeneracy of the energy levels. A great deal of effort has gone into the development of a detailed theory of the Stark broadening of lines in hydrogenic systems. For cooler stars, such as the Sun, there are also important effects of line broadening in heavier atoms due to collisions with hydrogen atoms. The study of these processes is difficult both for theoretical and experimental work and astronomers still use data which best fits the empirical analysis of their spectra.

In the 'fine analysis' of stellar spectra the assumption of LTE may be dropped. One then has the task of solving the hydrostatic equation which gives the relation between temperature and pressure in the atmosphere, the equation of radiative transfer which gives the radiation intensity I_v, and all of the equations from atomic physics for the excitation, ionization and recombination processes which determine the populations of the atomic energy levels. What makes the task really formidable is that the equations from atomic physics, which have to be solved for each point in the atmosphere, involve the local values of the temperature, density and intensity I_v which all depend on the structure of the atmosphere as a whole. Such work also requires, of course, that huge amounts of data should be provided by atomic physics. It turns out that departures from LTE are most important for the hot stars and non-LTE analyses have led to major revisions in He/H abundance ratios which are always of great interest to the cosmologists. The problems of cool stars are, however, no less formidable, since their atmospheres contain many molecules and their opacities can be completely dominated by absorption in molecular lines.

17.5. Nebula spectra

The production of spectra of gaseous nebulae often involves processes which are very simple, sometimes too simple to be easily studied in the laboratory! Thus some of the strongest emission lines are produced by radiation from metastable states which have lifetimes of several seconds, and in the laboratory such states are perturbed, before they can radiate, by other atoms or by the walls of a containing vessel. Such perturbations may not occur in the nebulae because the density is too low and there are no walls. Also many of the processes involve atomic hydrogen, which is the simplest atom of all, but is not so simple for laboratory experiments. The first observations of nebular spectra, by Huggins in 1864, produced a great surprise. The strongest lines could not be identified with anything seen in the laboratory and remained unidentified for a long time. Did the nebulae contain a new element, nebulium, unknown on Earth? The idea was not so unreasonable since the Sun contained a new element, called helium, which was only later identified in the laboratory. Nebulium only became unreasonable when the periodic table became sufficiently complete that there was no room for another element (unless it were a very heavy one) and the problem was finally solved by Bowen in 1927 who, using results from ultra-violet spectroscopy, identified the lines with the transition $2p^2\ ^1D \rightarrow 2p^2\ ^3P$ in doubly ionized oxygen. That was a good year for work to start on the theoretical interpretation. The lines could not be due to electric dipole radiation because initial and final states had the same parity and they are therefore referred to as forbidden lines. They could be due to electric dipole or magnetic quadrupole radiation allowing for departures from LS coupling (which could be estimated from the splitting of $2p^2\ ^3P_J$, $J = 0$, 1 and 2). In fact the $^1D-^3P\ O^{2+}$ lines are mainly due to magnetic dipole radiation but other lines in spectra of nebulae are due to electric quadrupole and even magnetic quadrupole transitions. The question is often asked 'why are the forbidden lines so strong in the spectra of nebulae?' and the answer often given is that there is no reason why not: the ion in the metastable state may have a long radiative lifetime but at the low densities in the nebulae it can stay in that state until it eventually does radiate. That, however, is only half of the story. The nebulae are produced by photo-ionization of hydrogen

$$H + h\nu \rightarrow H^+ + e$$

where the ionizing photons come from a hot star. The electron produced in such a process typically has an energy of several eV but the electron temperatures in the nebulae (determined from relative strength of spectral lines) are about 10^4 K corresponding to energies of less than 1 eV. Most of the energy of the colliding electrons goes into excitation processes such as

$$O^{2+}(^3P) + e \rightarrow O^{2+}(^1D) + e$$

(processes involving hydrogen are unimportant because its excited states are much higher, and in any case most of the hydrogen is ionized). The forbidden lines are therefore strong because they provide most of the energy loss.

In order to make quantitative interpretations of the spectra of nebulae one needs cross-sections for excitation of positive ions by electron impact. The first attempt to make such calculations was in a pioneering paper by Hebb and Menzel in 1939 but, due to their use of perturbation theory, they obtained results which were a good deal too large. All subsequent work has been based on 'close-coupling' approximation which uses an expansion of the type

$$\Psi = A \sum_i \psi_i(\mathbf{X})\theta_i(\mathbf{x})$$

where ψ_i is a wave function for a target state with co-ordinates $\mathbf{X}$, θ_i is a function for a colliding electron with co-ordinate $\mathbf{x}$, and A is an anti-symmetrization operator. This is, essentially, a generalization of Hartree–Fock theory to continuum states. One obtains a system of coupled integro-differential equations to be solved for the θ_i. A great deal of effort, much of it stimulated by problems in astronomy, has gone into obtaining accurate solutions of these equations. In all of this work one uses the Schrödinger picture of wave functions and space co-ordinates, since that is the picture best adapted to computation. Once accurate rates are known for all of the radiative and collisional processes one can interpret the observed spectra and deduce values of temperature and density and the relative abundances of the chemical elements.

The lines of hydrogen and helium observed in nebulae are produced by radiative captures,

$$H^+ + e \rightarrow H_n + h\nu,$$

followed by cascades. Observations of these lines provide one of the best methods of estimating He/H abundance ratios. In 'big bang' cosmologies it is assumed that H and He are primordial and that all of the heavier elements are produced by nuclear reactions in stars (I must confess to an unfashionable slight scepticism which arises from the fact that we never observe any 'primordial' objects). In all nebulae observed the spectra contain lines of H and He and also of heavier elements such as oxygen. Some of the nebulae, supposed to be most-nearly primordial, have low oxygen abundances and an estimate of the primordial He/H ratios is obtained on extrapolating to a zero (O/H) ratio.

The recombination lines of H and He are also observed by radio astronomers for transitions such as $(n+1) \rightarrow n$ with n large (several hundred). The populations of the high-n levels are determined by both radiative and collisional processes and the calculation of the required atomic data has involved clever use of the correspondence between classical and quantum mechanics. Because the mechanism is basically that of recombination, higher states are over-populated relative to lower states and interesting effects of maser action can occur.

17.6. The solar corona

Another problem in astronomy which has stimulated a great deal of work in atomic physics is that of interpreting the spectrum of the solar corona. The corona extends out to a distance of several solar radii and, before the days of space observations at ultra-violet and X-ray wavelengths, was most easily observed during eclipses of the Sun by the Moon. The decisive breakthrough was the identification by Edlén in 1942 of lines such as the coronal green line, which is due to Fe^{13+}, $3p\ ^2P_{3/2} \rightarrow 3p\ ^2P_{1/2}$. Such a highly ionized ion can, of course, only exist at high temperatures (10^6 K or more) but once such temperatures are accepted everything fits into place (the extent of the corona is, for example, consistent with use of the hydrostatic equation). The ionization is by collisions,

$$X + e \rightarrow X^+ + e + e,$$

and recombination is radiative,

$$X^+ + e \rightarrow X_n + h\nu.$$

The rates for both of these processes are proportional to electron density and the ionization equilibrium, $N(X^+)/N(X)$, depends only on temperature. Further progress requires knowledge of the collision cross-sections. The problem of collisional ionization remains one of the most difficult in atomic physics. We still do not have reliable methods for calculating wave functions for two relatively slow electrons in the field of a positive ion. Fairly good estimates of the collisional ionization rates can, nevertheless, be made, based on laboratory measurements, scaling laws from semi-classical theory and calculations which should give reasonably good results for highly-ionized systems. Some 25 years ago there was a worrying discrepancy between coronal temperatures deduced from studies of the ionization equilibrium on the one hand and of the line profiles (determined by the thermal Doppler effect) on the other. The problem was solved by Burgess who showed that the rate of recombination can be greatly enhanced by its going *via* compound resonance states,

$$X^+ + e \rightarrow X^{**} \rightarrow X_n + h\nu.$$

The resonances which occur in the process of *di-electronic* recombination are analogous to those which occur in the proton-capture reactions in stellar interiors but there is one important difference. Because the forces between X^+ and e^- is attractive, one can have an infinite number of resonances converging to the ionization limit and, at the high temperatures of the solar corona, allowance for the intermediate resonance states can increase the total recombination rate by as much as two orders of magnitude. More recent work has shown such Rydberg series of resonances are of great importance for practically all reactions, at thermal energies, involving electrons and positive ions.

18

Schrödinger's contribution to chemistry and biology

LINUS PAULING

Linus Pauling Institute of Science and Medicine, Palo Alto

My association with Erwin Schrödinger was not a close one, although I spent the summer of 1927 in Zürich, with the stated purpose (stated in my letters to the John Simon Guggenheim Memorial Foundation) of working under his supervision. In fact, I spent most of my time in my room, trying to solve the Schrödinger equation for a system consisting of two helium atoms. I did not have very much success, except that, as was mentioned later by John C. Slater, I formulated a determinant of the several spin-orbital functions of the individual electrons as a way of ensuring that the wave function is antisymmetric. This was a device that Slater made much use of in discussing the electronic structure of atoms and also of molecules in 1929 and 1931.

Walter Heitler and Fritz London were also in Zürich that summer, also with the plan of working with Schrödinger. They told me that they had talked with Schrödinger several times about their work, while walking through the woods. I did not have even that much contact with him, because he was working so hard on his own problems.

It might be possible to put theoretical physicists on a scale ranging from one extreme, those who deal with ideas, to the other, those who deal with mathematics. Wolfgang Pauli is an example of a theoretical physicist near the mathematics end. I remember that in August 1956, when he and I were participating in the Avogadro symposium in Rome, he said to me 'I have made only one non-mathematical discovery in my life, the discovery of the exclusion principle; and that is what I was given the Nobel prize for!' I remember also that at one time John H. Van Vleck said to me 'I have never made any discovery except by working with the equations,' and I replied 'I have never made any discovery except by having an idea, which I then tried to support by working with the equations.' Even though Schrödinger's fame rests upon his equation, I think that he should be placed near the idea-man end of the range. His great discovery was based upon an idea, the idea that the properties of atoms and molecules could be calculated by solution of a differential equation of a special sort, a characteristic-value equation. This idea and consideration of the classical equations of motion in the Hamiltonian form led him to the Schrödinger equation.

An example of an earlier idea of Schrödinger's is that in 1921 he treated many-electron atoms by idealizing the inner shells as having the electrons distributed uniformly over the surface of a sphere.

18.1 Chemistry

Sixty years ago, when quantum mechanics was discovered, the observed properties of chemical substances posed many puzzling questions, and those of biology still more. Quantum mechanics, especially in the form discovered by Schrödinger, introduced new concepts that made it possible to answer many of these questions, and as a result both chemistry and biology developed rapidly.

In 1916, when I first become interested in the question of the nature of the forces that hold atoms together in molecules and crystals, chemical bonding was discussed in terms of hooks and eyes that could link the atoms together. Valence was the number of hooks or eyes possessed by the atom. It was in 1916 that Gilbert Newton Lewis, in Berkeley, stated that the chemical bond consists of a pair of electrons held jointly by two atoms. He placed the eight electrons constituting a stable outer shell for many atoms at the corners of a cube, later in pairs at the corners of a tetrahedron. This idea seemed to contradict the Bohr model of atoms, in which the electrons move about the nucleus in orbits that were usually thought to keep them from being localized in the way described by Lewis, and chemists and physicists debated the apparent contradiction between the static atom and the dynamic atom. We can see now that Lewis's idea was not a bad one; the positions in which he placed the electrons are the average positions of the electrons in their orbits.

Even before quantum mechanics, theoretical physicists tackled the problem of the nature of the chemical bond. For example, both Heisenberg and Pauli tried to formulate a quantum theory of the hydrogen molecule ion, H_2^+, that would lead to agreement with the experimental value of the bond energy of its one-electron bond, but without success. Then quantum mechanics was developed, by Heisenberg, Dirac, and especially Schrödinger, and it became possible to attack problems of this sort.

In 1927 the young Danish physicist Ø. Burrau published his numerical solution of the Schrödinger wave equation for the hydrogen molecule. The value of the bond energy that he obtained was in good agreement with experiment, as was also the value of the bond length. The electron distribution function was shown to be concentrated around the two nuclei and the space between them. During the next year it was pointed out that Burrau's wave function could be approximated rather well simply by adding two functions, one in which the electron occupies the 1s orbital around one proton, and the other in which it occupies a similar orbital around the second proton. This wave function can be described as corresponding to resonance of the electron between the 1s orbitals around the two protons. The stability of the molecule – the bond energy – can be described as the resonance energy, in the same way as first introduced by Heisenberg in his treatment of the triplet states and singlet states of the helium atom. It was Heisenberg who introduced the word 'resonance' to describe this phenomenon of the decrease or increase of energy of a system when the wave functions for two equivalent structures are combined symmetrically or antisymmetrically. Heisenberg discussed the phenomenon in an especially effective way with use of two harmonic oscillators interacting through a weak coupling.

The development of the theory of the electron-pair bond was initiated in 1927 by the work of Condon and that of Heitler and London on the hydrogen molecule. Condon gave a discussion of the hydrogen molecule based upon Burrau's treatment of the hydrogen molecule ion. He introduced two electrons into the normal-state orbital described by Burrau for the one electron of the molecule ion, and calculated the bond energy of the hydrogen molecule with use of Burrau's energy values corrected by an estimate of the interaction energy of the two electrons with one another. Condon's treatment is the prototype of the molecular-orbital method of discussing the electronic structure of molecules. Heitler and London treated the hydrogen molecule by placing one electron, with positive spin, around the first nucleus, in its 1s orbital, and the second electron, with negative spin, around the other nucleus. The interaction energy calculated for this structure does not correspond to a significant bond. If, however, the wave functions for this structure and the structure with the electrons interchanged are added, the calculated interaction curve corresponds to the formation of a bond with bond energy approximately that observed. This treatment, in which the repulsion of these two electrons for one another is considered to be strong enough to keep them on different protons, is the prototype of the valence-bond method of discussing the electronic structure of molecules.

Several puzzling questions about chemical bonds were answered during the next few years, always in ways suggested by the Schrödinger wave equation. One question was that of the apparent equivalence of the four bonds of the carbon atom, as is indicated by the properties of carbon compounds. The four orbitals of the outer shell of the carbon atom, as given by the analysis of spectra and also by quantum mechanics, are of two kinds, one $2s$ orbital and three $2p$ orbitals. It was clear that, for the formation of a shared-electron-pair bond between two atoms, one needs an orbital of each atom and two electrons, one assigned to each orbital. The difference between the $2s$ orbital and the $2p$ orbitals would be expected to cause one of the bonds formed by a carbon atom to be different from the other three. In 1928 I pointed out that four hybrid orbitals could be formed, equivalent to one another, by linear combination of the wave functions for the $2s$ and $2p$ orbitals.

I shall not discuss the many other ways in which modern chemistry has developed through the introduction of ideas suggested by quantum mechanics. In addition to exploiting these ideas in a semi-empirical way, theoretical chemists have also developed better and better ways of obtaining approximate solutions of the Schrödinger equation for more and more complicated molecules. This procedure is especially important in the discussion of substances that cannot be subjected to experimental study or for which experimental studies are difficult. For smaller molecules, such as HNC or CH_2, the calculated values of energy of formation, bond lengths, bond angles, and other properties agree completely with available spectroscopic values, and for larger molecules the probable errors in the calculated values are getting smaller every year. There is no doubt that the Schrödinger equation provides the theoretical basis of chemistry.

18.2 Biology

During the last four decades biology has been revolutionized by the introduction of a new field, which is usually called molecular biology. The development of molecular biology has resulted almost entirely from the introduction of the new ideas into chemistry that were stimulated by quantum mechanics. It is accordingly justified, in my opinion, to say that Schrödinger, by formulating his wave equation, is basically responsible for modern biology.

We might now ask the following question: To what extent, aside from his discovery of the Schrödinger equation, did Schrödinger contribute to modern biology, to our understanding of the nature of life? It is my opinion that he did not make any contribution whatever, or that perhaps, by his discussion of 'negative entropy' in relation to life, he made a negative contribution.

In 1944 Schrödinger's book *What is Life?*, with the subtitle *The Physical Aspect of the Living Cell*, based on lectures given at Trinity College, Dublin, was published (Schrödinger, 1944). In this book he states that

'A living organism continually increases its entropy – or, as you may say, produces positive entropy – and thus tends to approach the dangerous state of maximum entropy, which is death. It can only keep aloof from it, i.e. alive, by continually drawing from its environment negative entropy – which is something very positive as we shall immediately see. What an organism feeds upon is negative entropy.'

Later, after a discussion of probability theory and entropy, he wrote

'How would we express in terms of the statistical theory the marvellous faculty of a living organism by which it delays the decay into thermodynamical equilibrium (death)? We said before: "It feeds upon negative entropy," attracting, as it were, a stream of negative entropy upon itself, to compensate the entropy increase it produces by living and thus to maintain itself on a stationary and fairly low entropy level.'

When I first read this book, over 40 years ago, I was disappointed. It was, and still is, my opinion that Schrödinger made no contribution to our understanding of life.

Several physicists and biologists with whom I have discussed this question have disagreed with me. When I asked what the contribution made by Schrödinger to our understanding of life is, each answered essentially by saying that Schrödinger showed that life is negative entropy, that living organisms utilize entropy in a way different from non-living matter.

I have not had the opportunity to discuss this matter with a physical chemist, a person with a good understanding of the great work by J. Willard Gibbs on chemical thermodynamics. I am sure that he or she would agree with me.

Schrödinger's discussion of thermodynamics is vague and superficial to an extent that should not be tolerated even in a popular lecture. In the discussion of thermodynamic quantities it is important to define the system. When he is writing about a change in entropy of the system, Schrödinger never defines the system. Sometimes he seems to consider that the system is a living organism with no interaction whatever with the environment; sometimes it is a living organism in thermal equilibrium with the environment; and sometimes it is the living organism plus the environment, that is the universe as a whole. In the first case a spontaneous process involves necessarily an increase in the entropy of the living organism; in the second case it involves an increase in entropy of the organism plus its environment, with the entropy of the organism either increasing or decreasing; and in the third case it involves an increase in the entropy of the universe.

In his note to Chapter 6 of *What is Life?* Schrödinger writes 'The remarks on *negative entropy* have met with doubt and opposition from physicist colleagues. Let me say first, that if I had been catering for them alone I should have let the discussion turn on free energy instead.'

Schrödinger might have discussed a poikilothermic organism encapsulated in such a way as to be in thermal equilibrium with its environment but not able to effect a material transfer through its capsule. The organism will continue to live for some time, undergoing various spontaneous reactions. We know from the principles of thermodynamics that the free energy of the organism necessarily decreases. One cannot say, without more information, whether there is an increase or a decrease in entropy of the organism.

Schrödinger was led to discuss entropy in relation to life through his learning that there is a genetic code. The general principles of heredity had been worked out, and it was recognized that there must be some sort of code, probably a linear arrangement of 'letters'. In 1943 the genetic material was usually considered to be protein, rather than nucleic acid, with the 'letters' the different amino-acid residues in a polypeptide chain, constituting the gene. It was known that proteins contain long polypeptide chains, sometimes with hundreds of amino-acid residues per chain. It was also already recognized that a particular protein, such as human hemoglobin, has a well-defined sequence of amino-acid residues in its polypeptide chains. A polypeptide chain with a random sequence of amino-acid residues would, other factors being neglected, have a larger entropy and hence a larger probability than one with a well-defined sequence. Schrödinger seems to have asked himself the question: 'What is the process that leads to the production of these well-defined polypeptide chains, with their low entropy?' He seems to have answered the question, in a rather vague way, by saying that the organism 'feeds upon negative entropy, attracting, as it were, a stream of negative entropy upon itself.'

The real question about the nature of life, which Schrödinger failed to recognize, is the question as to how biological specificity is achieved; that is, how the amino-acid residues are ordered into the well-defined sequence characteristic of the specific organism. The gene has the power to reproduce itself, and also to direct the synthesis of the protein for which it has responsibility.

There is no problem from the thermodynamic point of view. Let us assume that there is in the cell a reservoir of the various amino acids, in a situation such that condensation into polypeptide chains is accompanied by a decrease in free energy, whether the sequence is ordered or not.

It is the nature of enzymes that they have the ability to increase the rate of a reaction greatly, by many orders of magnitude, and also that they may be highly specific, able to distinguish one amino acid from another. It is not unreasonable that the gene might provide the information to permit a replica of itself to be synthesized and also to permit a particular polypeptide chain to be synthesized, with the specific reactions involved accelerated by a specific enzyme. The product, of course, has lower entropy than a similar product with a random sequence. It is accordingly something within the cell, rather than negative entropy from outside the cell, that is involved here.

In my own effort to understand the nature of life I have concentrated on the question of the interatomic or intermolecular interactions that can operate to lead to biological specificity, the detailed structures that confer individuality on organisms. The idea that the striking specificity shown in biological reactions results from the presence of mutually complementary combining regions on the interacting molecules was suggested in the period around 1930 by Breinl and Haurowitz, Jerome Alexander, and Stuart Mudd. There had been some intimation of it in the early work of Ehrlich and Bordet. If two molecules have such structures that a portion of the surface of one fits neatly against a portion of the surface of the other, then the electronic dispersion forces, formation of hydrogen bonds, and other weak intermolecular interactions might be great enough to lead to the formation of a bond between the two molecules that will resist the disrupting action of thermal agitation.

There is an alternative idea, that in some way a molecule has the power to direct the synthesis of a replica of itself. This was advanced in quantum mechanical form, as the explanation of biological specificity in serological reactions and the replication of genes, by the physicist Pascual Jordan, in 1940. He said that there is a stabilizing resonance interaction between two identical or nearly identical molecules or parts of molecules, such as to lead to the formation of a bond between them that would not be formed between unlike molecules. In 1940, Max Delbrück told me about Jordan's paper and three earlier ones that he had written on the same theme. I pointed out that the amount of resonance stabilization between identical molecules is in fact far too small, in relation to the energy of thermal agitation, to be effective in this way. I had in fact just written a paper about the structure of antibodies and the nature of the interactions between antibodies and antigens, and had reached the conclusion, later fully verified by experiment, that there is a detailed complementariness in structure between the antibody and the antigen. Later I suggested that the gene might consist of two mutually complementary chains, each of which could act as the template for the synthesis of the other. There is now no longer any doubt that the mechanism of biological specificity is detailed complementariness in structure of interacting molecules.

Schrödinger was, of course, right in pointing out that the existence of living organisms, which are highly improbable structures, depends upon our being in a universe that is far from equilibrium. We are so far from equilibrium that even highly improbable reactions can occur, without any violation of the laws of thermodynamics. Many of these highly improbable reactions depend upon having a seed, a template, that directs the reaction. Examples are known in inanimate nature. The responsible mechanism is the same as in living organisms, detailed molecular complementariness.

About 50 years ago an associate of mine in the California Institute of Technology synthesized an organic compound that had been discovered by an investigator in The Netherlands. When he determined the melting point, he found it to be lower than had been reported by the first investigator. He wrote to The Netherlands, asking for a sample of the compound, in order to compare it with his preparation. After the sample arrived, however, his own substance was found to have the higher melting point. The explanation is that there happened to be a seed in California that served as the center of crystallization of the unstable modification of the crystal. After the sample had arrived, there were seeds for the stable crystalline form of the substance, and from then on it was not possible to crystallize the unstable form.

The specificity shown in crystallization is astonishing. Crystals of potassium hydrogen tartrate often form in grape jelly. When these crystals are washed and analyzed, they are found to be essentially 100% pure. The crystal has selected from the hundreds or thousands of different kinds of molecules and ions in the grape jelly just the ones that it incorporates into its own structure. The growing crystal has a cavity on its surface that is closely complementary in structure to a tartrate ion, which fits into the cavity, whereas other ions and molecules do not fit and are rejected.

The discovery of the Schrödinger wave equation was a great event for physics. It may well be judged as having been an even greater event for chemistry and biology, which have progressed so rapidly during the last 60 years, with much of the progress owed to the new ideas provided by quantum mechanics. We are all indebted to Erwin Schrödinger for his great contribution.

Reference

Schrödinger, E. (1944) *What is Life? The Physical Aspect of the Living Cell.* Cambridge University Press

19

Erwin Schrödinger's What is Life? and molecular biology

M.F.PERUTZ

Medical Research Council Centre, Cambridge

In the early 1940s Schrödinger worked at the Institute for Advanced Studies in Dublin. One day he met P. P. Ewald, another German theoretician, then a professor at the University of Belfast. Ewald, who had been a student in Göttingen before the First World War, gave Schrödinger a paper that had been published in the *Nachrichten der Gesellschaft der Naturwissenschaften* in Göttingen in 1935 (Yoxen, 1979). The paper was by N. W. Timoféeff-Ressovsky, K. G. Zimmer and Max Delbrück (1935), and was entitled 'The nature of genetic mutations and the structure of the gene'. Apparently Schrödinger had been interested in that subject for some time, but the paper fascinated him so much that he made it the basis of a series of lectures at Trinity College, Dublin, in February 1943 and published them as a book in the following year, under the title *What is Life?* (Schrödinger, 1944). The book is written in an engaging, lively, almost poetic style (for example, 'The probable life time of a radioactive atom is less predictable than that of a healthy sparrow'). It aroused much interest, especially among young physicists, and helped to stimulate some of them to turn to biology. I was asked by the organizers of the Schrödinger Centenary Symposium to assess its significance for the development of molecular biology.

19.1. The Timoféeff-Ressovsky, Zimmer and Delbrück paper

This paper forms the basis of Schrödinger's book. It covers 55 pages and is divided into four sections. The first, by Timoféeff, describes the mutagenic effects of X-rays and γ-rays on the fruitfly *Drosophila melanogaster*. He shows that the spontaneous mutation rate of the fly is low, and that it is raised about five-fold by a rise in temperature of 10°C. Ionizing radiations increase that rate as a linear function of the dose, independent of its time distribution, of the wavelength and of the temperature during irradiation. The second section of the paper is by Zimmer and applies the target theory to Timoféeff's results. Since the number of mutations $x = a(1 - e^{-kD})$, where a and k are constants and D is the dose, no more than one hit can be needed to produce a mutation. (If more were needed then

$$x = a\left[1 - e^{-kD}\left(1 + kD + \frac{(kD)^2}{2!} + \frac{(kD)^3}{3!} + \cdots + \frac{(kD)^{n-1}}{(n-1)!}\right)\right] \quad .)$$

Zimmer next asks whether the mutations had arisen by the direct absorption of quanta, by the passage of secondary electrons through a sensitive volume, or by the generation of ion pairs. If the dose is measured in roentgens (r), the number of quanta required to produce a given dose diminishes with diminishing wavelength. Thus, direct absorption of quanta is inconsistent with the linear dependence of the mutation rate on the dose. The same applies to secondary electrons. Only the number of ion pairs is proportional to the dose, obviously, since that is how the dose is measured. Zimmer therefore concludes that a single hit suffices for the production of one mutation and that this hit consists in the formation of an ion pair or a transition to higher energy. He points out in a footnote that this conclusion applies only if the mean distance between two ionizations is large compared to the target. Five years later, H. J. Muller (1940) showed that this was in fact not true if hard radiation was used, but this point was proved to be irrelevant by later research, as we shall see.

The third section of the paper is by Delbrück and bears the title 'Atomphysikalisches Modell der Mutation' (a model of genetic mutation based on atomic physics). Delbrück reminds us that the concept of the gene began as an abstract one, independent of physics and chemistry, until it was linked to chromosomes and later to parts of chromosomes which were estimated to be of molecular size. The low natural mutation rate suggests that genes are stable and exceptionally inert to chemical attack by the reagents normally present in the living cell. The doubling of genes must be a property not just of the genes themselves, but of the genes working together with their surrounding substance. Since he and his colleagues had no means of discovering the chemical nature of genes directly, they attacked the problem indirectly by studying the nature and the limits of their stability and by asking if these are consistent with the knowledge that atomic theory has provided about the behaviour of well-defined assemblies of atoms.

Such assemblies can undergo the following discrete and spontaneous transitions:

(a) Vibrational. These are very frequent and involve no chemical changes.

(b) Electronic. From these the assemblies may either revert to the ground state or reach a new equilibrium state after undergoing an atomic rearrangement. The rate, W, of such an event is an exponential function of its activation energy U:

$$W = \frac{kT}{h} \exp -\frac{U}{kT},$$

where kT/h corresponds to the mean vibrational frequency of atoms and is about $10^{14}\,\mathrm{s}^{-1}$.

Delbrück gives an instructive table relating U in multiples of kT to $1/W$, the time when half the molecules will have undergone an electronic transition (see Table 19.1). These times range from $2 \times 10^{-10}\,\mathrm{s}$ for $U = 10kT$ to 30 000 years for $U = 60kT$; the corresponding activation energies range from 0.3 to 1.8 eV. Temperature coefficients range from 1.4 per 10 °C for $U = 10kT$ to 7.4 for 10 °C for $U = 60kT$. I have added another row to the table, showing that a carbon–carbon bond, whose cleavage requires an activation energy of about 3 eV, has a half life of 10^{30} years. Delbrück mentions that chemical bond energies are of the order of several electron volts, but argues that the activation energies of molecules cover an even wider range than his table, so that reaction rates of any order of magnitude can be expected for them. More definite conclusions can be

Table 19.1. *Relation between activation energy and probability of transitions*

$\dfrac{U}{kT}$	$W\ (\mathrm{s}^{-1})$	$\dfrac{1}{W}$	U (eV)	$\dfrac{W_{T+10}}{W_T}$
10	4.5×10^{9}	$2 \times 10^{-10}\,\mathrm{s}$	0.3	1.4
20	2.1×10^{5}	$5 \times 10^{-6}\,\mathrm{s}$	0.6	1.9
30	9.3	0.1 s	0.9	2.7
40	4.2×10^{-4}	33 min	1.2	3.8
50	1.9×10^{-8}	16 months	1.5	5.3
60	8.7×10^{-13}	30 000 years	1.8	7.4
116	10^{-38}	10^{30} years	3.0	84

U = activation energy; W = probability; eV = electron volts, where $1\,\mathrm{eV} = 23\,\mathrm{kcal}$; k = Boltzmann constant; T = temperature.

drawn by comparing the calculated temperature coefficients of the electronic transitions with those of the spontaneous mutations. If the mutation frequency rises about five-fold for a 10°C rise in temperature, then according to Delbrück's table this corresponds to an activation energy of ~ 1.5 eV and an average lifetime of a few years. Exceptionally frequent mutations needing lower activation energies were found to have lower temperature coefficients, exactly as theory predicts. In general, evolution had stabilized the molecular structure of genes so that their natural frequency of rearrangement is smaller by several orders of magnitude than the frequency of their reproduction. On the other hand, a single ionization or excitation should be sufficient to produce any given mutation, regardless of its natural frequency. This is also in accord with observation.

Delbrück then describes how, on average, X-rays lose energy to secondary electrons in portions of 30 eV per ionization, which is $1000kT$ and 20 times the energy of activation of 1.5 eV needed for a spontaneous mutation. However, to produce as much as 1.5 eV, the ionization must not occur too far away from its target. He knew too little about the ways in which the energy of photoelectrons is dissipated to determine the absolute value of the dose needed to induce a mutation with a probability of unity, but it was expected that the dose, expressed as the number of ionizations per unit volume, would be about 10–100 times smaller than the number of atoms of the gene per unit volume. Delbrück now calculates that dose as follows.

A frequently observed X-ray mutation (eosin) occurs with a dose of 6000 r once in 7000 gametes. Hence a probability of unity of its occurrence needs a dose of 42×10^6 r. 1 r produces $\sim 2 \times 10^{12}$ ion pairs in 1 ml H_2O, whence 42×10^6 r produce $\sim 10^{20}$ ion pairs. Since 1 ml H_2O contains $\sim 10^{23}$ atoms, this means that at least one in a thousand atoms becomes ionized. However, Delbrück cautiously refrains from concluding that a gene is likely to consist of a thousand atoms.

Schrödinger used this result to point out 'that there is a fair chance of producing a mutation when an ionization occurs not more than about 10 atoms away from a particular spot on the chromosome', but research published while Schrödinger was writing his book showed such calculations to be meaningless. In a paper that appeared in *Nature* in June 1944, Joseph Weiss (1944) pointed out that the biological effects of ionizing radiation are due principally to the generation of hydroxyl radicals and hydrogen atoms in the surrounding water. Collinson *et al.* (1962), and independently Czapski and Schwartz (1962), later discovered that the hydrogen atoms were in fact hydrated electrons (Dainton, 1975). Hydroxyl radicals and hydrated electrons have half lives of $\sim 1\,ms$ (assuming a concentration of $1\,\mu$mole H_2O_2) and $0.5\,ms$, respectively, in which time they can diffuse to their targets even if they are generated more than a thousand atomic diameters away from them.

Delbrück concluded that mutations are quantum transitions due either to random thermal fluctuations or to the absorption of radiant energy; these transitions induce either a rearrangement of the atomic assembly to a different state of equilibrium (such as a tautomeric change from maleic to fumaric acid) or the cleavage of a chemical bond. Spontaneous mutations arise predominantly from thermal fluctuations rather than from natural radiation, and they are independent of the stage of reproduction of the gene, but there exist also mutable alleles in which mutations do arise during reproduction. Delbrück predicts correctly that ultraviolet light may induce mutations. He argues that mutations found to be produced by sudden heat shock look as though they were not induced by the influence of temperature on the mutation rate as such, but by energy-generating defence mechanisms induced in the organisms. Here again he was right.

Delbrück argues that it is premature to make the description of the gene any more concrete than the following:

'We leave open the question whether the single gene is a polymeric entity that arises by the repetition of identical atomic structures or whether such periodicity is absent; and whether individual genes are separate atomic assemblies or largely autonomous parts of a large structure, i.e. whether a chromosome contains a row of separate genes like a string of pearls, or a physico-chemical continuum.'

I found the Timoféeff–Zimmer–Delbrück papers, and especially Delbrück's part, most impressive. His interest in biology had been aroused by a lecture of Niels Bohr on 'Light and life' delivered in Copenhagen in 1932. In that lecture Bohr had said:

'The existence of life must be considered as an elementary fact that cannot be explained, but must be taken as a starting point in biology, in a similar way as the quantum of action, which appears as an irrational element from the point of view of classical mechanical physics, taken together with the existence of elementary particles, forms the foundation of atomic physics. The asserted impossibility of a physical or chemical explanation of the function peculiar to life would be ... analogous to the insufficiency of the mechanical analysis for the understanding of the stability of atoms.'

See Bohr (1933). The search for Bohr's elementary fact of life fired Delbrück's imagination. He was only 29 years old, working as assistant to Otto Hahn and Lise Meitner in the Kaiser Wilhelm Institut für Chemie in Berlin and doing his biological work as a side line, but his paper shows the maturity, judgement and breadth of knowledge of someone who had been in the field for years. It is imaginative and sober, and its carefully worded predictions have stood the test of time. The paper won him a Rockefeller Fellowship to Pasadena to work with the *Drosophila* geneticist T. H. Morgan. There he met Linus Pauling with whom he published an important paper in 1940. That paper was an attack on the German theoretician P. Jordan who had advanced the idea that there exists a quantum-mechanical stabilizing interaction, operating preferentially between identical or near identical molecules, which is important in biological processes such as the reproduction of genes. Pauling and Delbrück pointed out that interactions between molecules were now rather well understood and gave stability to two molecules of *complementary* structure in juxtaposition, rather than to two molecules with necessarily *identical* structures. Complementariness should be given primary consideration in the discussion of the specific attraction between molecules and their enzymatic synthesis (Pauling and Delbrück, 1940). In 1937 the Cambridge geneticist and biochemist J. B. S. Haldane had made a similar suggestion: 'We could conceive of a [copying] process [of the gene] analogous to the copying of a gramophone record by the intermediation of a negative, perhaps related to the original as an antibody is to an antigen' (Haldane, 1937). Schrödinger mentions neither of these important suggestions.

19.2. Erwin Schrödinger's *What is Life?*

Schrödinger's book is written for the layman from the physicist's point of view, and begins with a chapter on 'The classical physicist's approach to the subject'. He asks how events in space and time taking place in a living organism can be accounted for by physics and chemistry.

'Enough is known about the material structure of life to tell exactly why present-day physics cannot account for life. That difference lies in the statistical point of view. It is well-nigh unthinkable that the laws and regulations thus discovered (i.e. by physics) should apply immediately to the behaviour of systems which do not exhibit the structure on which these laws and regularities are based.'

Schrödinger jumps to that conclusion after reading that genes are specific molecules of which each cell generally contains no more than two copies. He had entered Vienna University in 1906, the year that Boltzmann died, and had been taught physics by Boltzmann's pupils. He remained deeply influenced by Boltzmann's thoughts throughout his life. According to Boltzmann's statistical thermodynamics, the behaviour of single molecules is unpredictable; only the behaviour of large numbers is predictable. In genetics, therefore, Schrödinger concludes 'we are faced with a mechanism entirely different from the probabilistic ones of physics'. This difference forms the guiding theme of his book.

In the first chapter, Schrödinger illustrates the meaning of statistical thermodynamics by the examples of Curie's law, of Brownian motion and diffusion, and of the $\sqrt{n}$ rule. His next two chapters, on the 'Hereditary mechanisms' and on 'Mutations', give brief popular introductions to text book knowledge on these subjects available at the time. They reveal one vital misconception in Schrödinger's mind: 'Chromosomes', he writes, 'are both the law code and the executive power of the living cell'. In fact, biochemists had shown that the executive power resides in enzyme catalysts, and G. W. Beadle and E. L. Tatum (1941) discovered that single genes determine single enzymatic activities; that discovery led to the 'one gene one enzyme hypothesis', an idea that had already been foreshadowed by Haldane (1937). Schrödinger does not appear to have heard of this.

His next two chapters that form the backbone of his book: 'The quantum-mechanical evidence' and 'Delbrück's model discussed and tested' are based entirely on the Timoféeff–Zimmer–Delbrück papers. In fact they are largely paraphrased versions of them; even Schrödinger's famous hypothesis that the gene is like an aperiodic one-dimensional crystal is a reformulation of Delbrück's suggestion that 'the gene is a polymer that arises by the repetition of identical atomic structures'. One could argue over the distinction between aperiodic and identical, but Delbrück could not have meant structures that are *completely* identical, since these could contain no information. To my mind the two hypotheses mean the same.

The last two chapters do contain Schrödinger's own thoughts on the nature of life. In his chapter on 'Order, disorder and entropy' he argues that 'the living organism seems to be a macroscopic system which in part of its behaviour approaches to that purely mechanical (as contrasted with thermodynamical) conduct to which all systems tend, as the temperature approaches the absolute zero and the molecular disorder is removed'. He comes to this strange conclusion on the ground that living systems do not come to thermodynamic equilibrium, defined as the state of maximum entropy. They avoid doing so, according to Schrödinger, by feeding on negative entropy. I suspect that Schrödinger got that idea from a lecture by Boltzmann on the second law of thermodynamics, delivered before the Imperial Austrian Academy of Sciences in 1886 (Boltzmann, 1886).

'Der allgemeine Daseinskampf der Lebewesen ist daher nicht ein Kampf um die Grundstoffe – die Grundstoffe aller Organismen sind in Luft, Wasser und Erdboden im Überflusse vorhanden – auch nicht um Energie, welche in Form von Wärme leider unverwandelbar in jedem Körper reichlich enthalten ist, sondern ein Kampf um die Entropie, welche durch den Übergang der Energie von der heissen Sonne zur kalten Erde disponibel wird.'
[Hence the general battle for existence of living organisms is not one for the basic substances – these substances are abundant in the air, in water and on the ground – also not for energy that every body contains abundantly in the form of heat, though unfortunately in a non-available form, but for entropy which becomes available by the transition of energy from the hot sun to the cold earth.]

Franz (later Sir Francis) Simon, then at Oxford, pointed out to Schrödinger that we do not live on $-T\Delta S$ alone, but on free energy. Schrödinger deals with that objection in the second edition of his book; he writes that he had realized the importance of free energy but had regarded it as too difficult a term for his lay audience; to me this seems a strange argument, since the meaning of entropy is surely harder to grasp. Schrödinger's postscript did not satisfy Simon who pointed out to him that

'The reactions in the living body are only partly reversible and consequently heat is developed of which we have to get rid to the surroundings. With this irreversibly produced heat also flow small amounts (either $+$ or $-$) of reversibly produced heat ($T\Delta S$), but they are quite insignificant and therefore cannot have the important effects on life processes which you assign to them.'

See Yoxen (1979). In fact, it was known when Schrödinger wrote his book that the primary currency of chemical energy in living cells is ATP (adenosine triphosphate), and that the free energy stored in ATP is predominantly enthalpic. However, Schrödinger did not remove this misleading chapter from later editions.

We now come to his final chapter entitled 'Is life based on the laws of physics?'. This chapter reiterates and amplifies the central argument already stated at the beginning of the book. According to Delbrück, he writes, the gene is a molecule, but the bond energies in molecules are of the same order as the energy between atoms in solids, for example in crystals, where the same pattern is repeated periodically in three dimensions, and where there exists a continuity of chemical bonds extending over large distances. This leads him to suggest that the gene is a linear one-dimensional crystal, but lacking a periodic repeat: an aperiodic crystal. I find it strange that he does not call it a polymer, as Delbrück did.

A single such crystal, or a pair of them, directs the orderly processes of life.

Yet, according to Boltzmann's laws, their behaviour must be unpredictably erratic. Schrödinger concludes that

'We are faced with a mechanism entirely different from the probabilistic one of physics, one that cannot be reduced to the ordinary laws of physics, not on the ground that there is any "new force" directing the behaviour of single atoms within an organism, but because the construction is different from any yet tested in the physical laboratory.'

Schrödinger argues that this allows only one general conclusion:

'Living matter, while not eluding the laws of physics as established to date, is likely to involve other laws of physics hitherto unknown which, however, once they have been revealed, will form as integral a part of this science as the former.'

Schrödinger is thus drawn to the same conclusion as Niels Bohr had been, apparently unknown to Schrödinger, 12 years earlier, and one that young physicists found equally inspiring.

Schrödinger next refers to a paper by Max Planck on dynamical and statistical laws. Dynamical laws control large scale events like the motions of the planets or of clocks. Clockworks function dynamically, because they are made of solids kept in shape by London–Heitler forces strong enough to elude disorderly heat motions at ordinary temperatures. An organism is like a clockwork in that it also hinges upon a solid: the aperiodic crystal forming the hereditary substance, largely withdrawn from the disorder of heat motion. The single cog of this clockwork is not of coarse human make, but is the finest piece ever achieved along the lines of the Lord's quantum mechanics. C. D. Darlington at Oxford had advised him that genes are likely to be protein molecules, as was then generally believed; Schrödinger quotes that, but does not mention that proteins are long chain polymers made up of some 20 different links that might form the kind of aperiodic patterns he had in mind. He must also have been unaware that the true chemical nature of that 'finest piece' was actually published while he was writing his book. In January 1944 there appeared in the *Journal of Experimental Medicine* a paper by O. T. Avery, C. M. McLeod and M. McCarty (1944) which reported conclusive evidence that genes are made not of protein, but of DNA (deoxynucleic acid). In the fullness of time, that discovery has led the majority of scientists to the recognition that life can be explained on the basis of the existing laws of physics.

Schrödinger's book made a considerable impact; but for it the Timoféeff, Zimmer and Delbrück paper that was published in an obscure journal would have remained unknown. Without *What is Life?* three prominent molecular biologists, Gunther Stent, Seymour Benzer and Maurice

Wilkins might not have entered the field. Sadly, however, a close study of his book and of the related literature has shown me that what was true in his book was not original, and most of what was original was known not to be true even when the book was written. Moreover, the book ignores some crucial discoveries that were published before it went into print. On the other hand, my reading has raised even further my already great respect and admiration for Delbrück's analytical powers and scientific rigour, and for the prophetic and imaginative concepts formulated by Haldane and Pauling, often long in advance of the relevant discoveries.

19.3. Replication of genes and protein synthesis as statistical mechanisms

The apparent contradictions between life and the statistical laws of physics can be resolved by invoking a science largely ignored by Schrödinger. That science is chemistry. When Schrödinger wrote 'The regular course of events, governed by the laws of physics, is never the consequence of one well-ordered configuration of atoms, not unless that configuration repeats itself many times', he failed to realize that this is exactly how chemical catalysts work. Given a source of free energy, a well-ordered configuration of atoms in a single molecule of an enzyme catalyst can direct the formation of an ordered stereospecific compound at a rate of 10^3-10^5 molecules a second, thus creating order from disorder at the ultimate expense of solar energy.

Chemists could also have told him that there is no problem in explaining the stability of polymers that living matter is made of, because their bond energies range from 3 eV upwards which corresponds to a half life for each bond of at least 10^{30} years at room temperature. The difficulty resides in explaining how their aperiodic patterns are accurately reproduced in each generation. There is no mention of this central problem in Schrödinger's book.

Let me now explain how research has cleared away the apparent contradiction between the randomness of single molecular events and the orderliness of life that exercised Schrödinger. The orderliness depends on fidelity of reproduction of the genetic message every time a cell divides, and of protein synthesis.

The genetic message is encoded in a sequence of nucleotide bases along a chain of DNA. That chain is paired to another carrying a complementary sequence of bases. The two chains are coiled around each other in a double helix in which each adenine (A) forms two hydrogen bonds with a thymine (T) and each guanine (G) forms three hydrogen bonds with a cytidine (C).

At cell division the two strands of the parent double helix separate and each forms a template for the formation of a new complementary strand, resulting in two daughter double helices with the same base sequence as the parent double helix. The necessary monomers are supplied in the form of nucleoside triphosphates which carry the energy for the formation of the growing chain in the form of an energy-rich P–O bond; the synthesis of new chain links is catalyzed by an enzyme or system of enzymes that attach themselves to the end of the double helix, unwind it, hold each parent strand rigidly in the conformation needed to catalyze the formation of a new chain link, move forward one step, catalyze the formation of the next link, and so on. The enzymes arrest the random motions of the DNA chain, thus allowing an orderly process to take place in a single molecule carrying the genetic information. Arthur Kornberg and his colleagues at Stanford University have worked out how these enzymes function in *Escherichia coli* (Kornberg, 1980, 1982).

How do they ensure that at each step of elongation only the nucleotide complementary to that on the parent strand is linked to the daughter strand? Chemical kinetics tell us that the four alternative trinucleotides must be bombarding the active site of the enzyme at the diffusion rate of about 10^9 molecules a second. On the other hand, their rates of dissociation from the active site vary, depending on their ability of forming complementary hydrogen bonds with the base of the parent strand. Only if the incoming nucleoside triphosphate is oriented correctly in the active site of the enzyme and if the hydrogen bonding groups of its base are complementary to those of the parent base will the new nucleotide remain in the active site long enough for a new chain link to be formed.

As Delbrück foresaw, the main source of spontaneous mutations is not the cleavage of covalent bonds in the parent strand. One source was believed to be the existence of tautomeric forms of the bases which differ in their arrangement of hydrogen bond donor and acceptor groups. Such tautomeric changes would allow a G to pair with a T, or a C with an A, or a G with an A, but in fact such mispairing probably occurs by another mechanism at apparently lesser cost in free energy. X-ray analysis of synthetic oligonucleotides has shown that mismatched bases can form hydrogen bonds with each other and be incorporated in the double helix with only slight distortions of the bond angles in the phosphate ester chain. Finally, a G–A pair can form, again with only minor distortions of the double helix, if either of the two bases is inverted about its bond to the ribose (Kennard, 1987). Judging by the frequency of these mistakes, the error rate in the reproduction of the genetic message should be 10^{-4} to 10^{-5} per nucleotide; in fact the measured error rate in *E. coli* is 10^{-8} to 10^{-10}, at least three orders of magnitude less than the expected one. How does Nature defeat statistical thermodynamics? One of its tricks is a proof-reading and editing mechanism, unravelled, initially by Kornberg and others at Stanford, and subsequently by A. R. Fersht and his colleagues, first in Stanford and later at Imperial College (Fersht, 1981; Kornberg, 1980, 1982). In *E. coli* the enzyme that catalyzes the elongation of the DNA chain has a 'second look' at the base pair just joined to the daughter double helix; it excises wrongly paired bases, followed by incorporation of the correct ones. Proof-reading and editing, however, is also subject to the errors imposed by chemical kinetics, which means that once in a thousand times, say, the correct base is excised and must then be re-incorporated in the growing chain. This costs energy. If proof-reading is too rigorous, too much energy is wasted in excising and re-incorporating correctly paired bases; if it is not rigorous enough, too many copying errors are left uncorrected. Using bacterial mutants which are either exceptionally error-prone or error-free, Fersht has measured the cost of fidelity and shown how Nature achieves the best compromise by increasing fidelity by two orders of magnitude to about 5×10^{-7}, but still not enough to account for the observed mutation frequency of only 10^{-8}–10^{-10} in the replication of *E. coli*. Glickman and Radman (1980) have discovered a second proof-reading mechanism that can distinguish the parent strand from the daughter strand by virtue of the fact that some of the parent strand's bases have become methylated, while those of the daughter strand are still bare. When it finds a mismatched base pair in the daughter strand, it excises it and replaces it by the correct one, thus reducing the error rate by the missing one or two orders of magnitude. (For a review of mismatch repair in *E. coli* see Radman and Wagner, 1986). The error rate of a viral RNA by an enzyme that is incapable of repairing mismatches was found to be about 10^{-4} per doubling, a quite unacceptably high rate even for a bacterium (Batschelet, Domingo and Weissmann, 1976).

The genetic message's function is to code for the sequence of amino acids along protein chains, but DNA does not code for proteins directly. Instead, the genetic message is first transcribed into messenger RNA and then translated into a sequence of amino acids in a protein chain. If enzymes are to work effectively, mistakes in the sequence must be infrequent. Transcription of DNA into RNA is not subjected to proof-reading and excision repair, perhaps because messenger RNAs are rarely longer than 10^4 base pairs, so that greater error rates are acceptable than in DNA replication. On the other hand, translation of RNA into protein presents problems which were first pointed out by Pauling, characteristically, *before* the enzymatic machinery for protein synthesis was unravelled (Pauling, 1958). Certain pairs of amino acids differ only by one methyl group. It is easy to imagine an active site of an enzyme that efficiently rejects an amino acid that is a misfit because it is too large by one methyl group, but it is hard to see how an active site could discriminate well against an amino acid that merely leaves a hole, because it is too short by one methyl group. The ratio of the reaction rates v of two amino acids A and B whose side chains differ in length by a single methyl group, one just fitting the active site and the other being too short, is given by the equation

$$v_A/v_B = \frac{[A]}{[B]} e^{-\Delta G_b/RT},$$

where ΔG_b is the difference in Gibbs binding energy due to the contribution of the side chain. ΔG_b is not likely to be more than $3 \, kcal \, mole^{-1}$. If $[A] = [B]$ this means that $v_A/v_B < 200$, implying an error rate greater than 0.5% (Fersht, 1981). Yet when Loftfield and Vanderjagt (1972) tried to measure the error rate in such a situation, they found it to be only three parts in 10 000, and concluded 'that the precision . . . of peptide assembly is very great, much greater than can be deduced from the study of non-biological chemical reactions'. Fersht (1981) showed how this low error rate is achieved without making Boltzmann turn in his grave. Nature makes use of the fact that the selection of the correct amino acid into the growing protein chain proceeds in two stages, both catalyzed by the same enzyme. In the first stage the amino acid is coupled to a phosphate to give it an energy-rich bond; in the second stage it is transferred to an adaptor, a molecule of RNA that carries the anticodon triplet of nucleotide bases complementary to the coding triplet for that particular amino acid. At the first stage of the reaction, the enzyme rejects amino acids with side chains that are misfits in the active site because they are too long, but reacts with amino acids whose side chains are too short with the large error rates predicted by Pauling. The second stage of the reaction takes place at a different active site of the same enzyme. It is constructed so as to fit those amino acids which were too short for the previous active site. It cleaves them from the adaptor RNA and sets them free some hundreds of times faster than the correct amino acid. This second stage can thus reduce the error by a further two orders of magnitude, giving a total error rate of only 10^{-4}. Fersht calls this a double sieve mechanism: the first sieve rejects amino acids that are too large; the second rejects the ones that are too small. A further stage of editing may reduce possible errors in the recognition by the coding triplet on the messenger RNA of the anticodon triplet on the adaptor RNA, thus ensuring incorporation of the correct amino acid into the growing protein chain (Hopfield, 1974; Thompson and Stone, 1977; Yates, 1979).

We can see that life has resolved the apparent conflict between the unpredictable behaviour of single molecules and the need for order by making enzymes sufficiently large to stabilize them in unique structures, capable of immobilizing other molecules in their active sites and bringing them into juxtaposition so that they can react at high rates. Yet enzymes are long chain polymers. What makes their chains fold up to form unique and largely rigid structures when entropy drives them to form random coils? X-ray analysis has shown the interior of proteins to be closely packed jig-saws of amino acids with hydrocarbon side chains that adhere to each other. They do so partly by dispersion forces, which are enthalpic, and partly thanks to the entropy that is gained by the exclusion of water from the protein interior. When the protein chain achieves maximum entropy by unfolding to a random coil, both polar and non-polar groups are exposed to water which adheres to them and becomes immobilized so that its entropy is diminished. When the chain folds up to its unique structure, the polar groups on the main chain form hydrogen bonds with each other, the non-polar side chains pack together and the bound water molecules are set free. The resulting gain in translational and rotational entropy of the water more than compensates for the loss of rotational entropy of the protein chain. Thus it is the water molecules' anarchic distaste for the orderly regimentation imposed upon them by the unfolded protein chain that provides a major part of the stabilizing free energy of the folded one and keeps it in its unique, enzymatically active structure.

19.4. Postscript

I feel that I should not close this story without telling what became of the scientists whose paper Schrödinger popularized.

Delbrück, whose Rockefeller Fellowship had taken him to Pasadena, remained there, with short interruptions, for the remainder of his life. In the early 1940s, he founded bacteriophage genetics and later, together with S. Luria, bacterial genetics; he became the leader of an enthusiastic band of young people who developed these new fields of research. In 1969 he, Luria and A. D. Hershey received the Nobel Prize for medicine or physiology 'for their discoveries concerning the replication mechanism and the genetic structure of viruses'. Delbrück died in Pasadena in 1981. I recently sketched his biography (Perutz, 1986).

While Delbrück had a happy life, Timoféeff's seems tragic to me, even though I am told that he did not regard it as such. He began his research on *Drosophila* in Moscow in the early 1920s. According to Zhores Medvedev (1982)

'In 1924, the Soviet Government made a special exchange agreement with Germany. The famous Kaiser Wilhelm Institute for Brain Research in Berlin-Buch was invited to help organize in Moscow a laboratory for brain research, specially designed to study the brain of Lenin, who had died in January 1924. (At the time of his death, Lenin was considered to be the greatest of geniuses, and his brain was expected to be unique. No subsequent publications in this special field of brain research were made by Soviet scientists.) In exchange, the Soviet Academy of Sciences promised to help set up in Berlin a laboratory of experimental genetics. Among the young scientists who were recommended to start work in the Berlin laboratory was Timoféeff. He left for Germany in 1926.'

In the 1930s Timoféeff thought of returning to Russia, but his friends advised him that it would be unsafe because Stalin's persecution of geneticists had already begun. His younger brothers were arrested and one was executed. In Germany during the war his elder son joined an underground antifascist group and was caught by the Gestapo and disappeared, even though he was no more than 17 years old. After the war had ended, in August 1945, Timoféeff himself was arrested by the Soviet secret police, sentenced to ten years' hard labour and sent to a prison camp in North Kazakhstan. Later he shared a prison cell with Alexander Solzhenitsyn in Bytyrsky with 22 other prisoners. Solzhenitsyn describes in the *Gulag Archipelago* how Timoféeff's irrepressible enthusiasm for science made him organize scientific seminars even in that prison cell. Apparently Solzhenitsyn also used him as a model for the scientist in *The First Circle*.

In 1947 Frédéric Joliot-Curie wrote to Beria, the head of the Russian secret police, on behalf of the French Academy of Sciences and asked for Timoféeff's release on the grounds that he was a valuable scientist and should be given a chance to do research. Joliot-Curie's intervention was successful. After several months of recuperation in a Moscow hospital, Timoféeff was sufficiently restored to set up a new secret prison research institute on radiation biology east of the Urals.

Previously, in September 1945, the Russians had also arrested Karl G. Zimmer with two of his colleagues in Berlin and had taken them to the Lubljanka Prison in Moscow for interrogation. After some time they were sent to do administrative work in a uranium factory not far from Moscow. When Timoféeff set up his new institute, he asked that Zimmer and his colleagues, as well as his own wife, should be allowed to join him there.

Timoféeff's eyesight had been ruined by starvation and his wife read the scientific literature to him. After Stalin's death they were released from prison and continued their work in Sverdlovsk. In 1964 Timoféeff was asked to organize the Department of Genetics and Radiobiology at the new Institute of Medical Radiology in Obninsk where Medvedev, the geneticist and author of the famous book *The Rise and Fall of Lysenko*, joined him. Medvedev describes him as a great man and a brilliant scientist; his mastery of many fields of genetics and biology, his dynamism and personal magnetism stimulated the work of the entire laboratory. Peter Kapitsa became his close friend.

Timoféeff retired in 1970 on a pension so meagre that it left him almost destitute. He died in 1981, in the same year as his friend Delbrück, who actually came to visit him in Obninsk after having been given the Nobel Prize in Stockholm. But for Schrödinger's book his name would have remained unknown outside the circles of genetics and radiation biology.

Zimmer returned to West Germany in 1955. He became one of the first to recognize the importance of electron spin resonance for radiation biology and to prove that ionizing radiations generate free radicals in biological molecules. In 1957, he was offered a chair in Heidelberg combined with a new Department of Radiation Biology at the Institute for Nuclear Research in Karlsruhe. There he worked on the effects of ionizing radiations on DNA and other biologically important molecules, and his laboratory became a successful centre for fundamental and applied radiobiology. He also published a book on that subject (Zimmer, 1960). He now lives in retirement in Karlsruhe.

As a final irony Traut (1963), working in Zimmer's laboratory, found the linear dose-response curve of Timoféeff's to have been an artefact. He showed that the mutation rate of *Drosophila* germ cells varies widely at different stages of their development. If males are irradiated and then mated, the frequency of mutation among the offspring varies with the time that has elapsed between the two events, because the sperm that fertilizes a female five days after irradiation was at an earlier stage of development when it was irradiated than the sperm that fertilizes a female one day after irradiation. At all stages the dose-response curves are non-linear. Traut demonstrated that a linear response curve, similar to those observed by Timoféeff, is obtained by summing the different dose-response curves produced by matings during the first four days of irradiation.

Zimmer (1966) comments:

This result removes one of the foundation-stones of the Green Pamphlet (as the Timoféeff, Zimmer & Delbrück paper became known). Strangely enough, that does not seem to matter any more, for two reasons; (i) the concept of the gene and modern trends in genetic research as well as in radiation biology have changed considerably during thirty years, and (ii) the Green Pamphlet has served a useful purpose by helping to initiate these modern trends.

References

Avery, O. T., McLeod, C. M. and McCarty, M. (1944) *J. Exp. Med.* 79, 137–58

Batschelet, E., Domingo, E. and Weissmann, C. (1976) *Gene* 1, 27–33

Beadle, G. W. and Tatum, E. L. (1941) *Proc. Natl. Acad. Sci. USA* 27, 499–506

Bohr, N. (1933) *Nature* 131, 458–60

Boltzmann, L. (1886) Der zweite Hauptsatz der mechanischen Wärmetheorie. *Sitzungsber. Kaiserl. Akad. Wiss. Wien*

Collinson, E., Dainton, F. S., Smith, D. R. and Tazuke, S. (1962) *Proc. Chem. Soc.* 140–4

Czapski, G. and Schwartz, H. A. (1962) *J. Phys. Chem.* 66, 471–9

Dainton, F. S. (1975) *Chem. Soc. Rev.* 4, 323–62

Fersht, A. R. (1981) *Proc. Roy. Soc.* B 212, 351–79

Glickman, B. W. and Radman, M. (1980) *Proc. Natl. Acad. Sci. USA* 77, 1063–7

Haldane, J. B. S. (1937) *The Biochemistry of the Individual* in *Perspectives of Biochemistry* (Needham, J. and Green, D. E., eds), pp. 1–10. Cambridge University Press

Hopfield, J. J. (1974) *Proc. Natl. Acad. Sci. USA* 77, 4135–9

Kennard, O. (1987) *Trends Nuc. Acad Res.* (in press)

Kornberg, A. (1980) *DNA Replication*. W. H. Freeman & Co., San Francisco

Kornberg, A. (1982) *Supplement to DNA Replication*. W. H. Freeman & Co., San Francisco

Loftfield, R. B. and Vanderjagt, D. (1972) *Biochem. J.* 128, 1353–6

Medvedev, Z. A. (1982) *Genetics* 100, 1–5

Muller, H. J. (1940) *J. Genetics* 40, 1–66

Pauling, L. (1958) *Festschrift Prof. Dr. Arthur Stoll Siebzigsten Geburtstag*, pp. 597–602

Pauling, L. and Delbrück, M. (1940) *Science* 92, 77–9

Perutz, M. F. (1986) *Nature* 320, 639–40

Radman, M. and Wagner, R. (1986) *Ann. Rev. Genet.* 20, 523–38

Schrödinger, E. (1944) *What is Life? The Physical Aspect of the Living Cell.* Cambridge University Press

Thompson, R. C. and Stone, P. J. (1977) *Proc. Natl. Acad. Sci. USA* 74, 198–202

Traut, H. (1963) *Dose-dependence of the Frequency of Radiation-induced Recessive Sex-linked Lethals in* Drosophila melanogaster, *with Special Consideration of the Stage Sensitivity of the Irradiated Germ Cells* in *Repair from Genetic Radiation Damage* (Sobels, F. D., ed.), p. 359. Pergamon Press, London

Timoféeff-Ressovsky, N. W., Zimmer, K. G. and Delbrück, M. (1935) *Nachrichten aus der Biologie der Gesellschaft der Wissenschaften Göttingen* 1, 189–245

Weiss, J. (1944) *Nature* 153, 748–50

Yates, J. L. (1979) *J. Biol. Chem.* 254, 1150–4

Yoxen, E. J. (1979) *Hist. Sci.* 17, 17–52

Zimmer, K. G. (1960) *Studien zur quantitativen Strahlenbiologie.* Wiesbaden

Zimmer, K. G. (1966) *The Target Theory* in *Phage and the Origins of Molecular Biology* (Cairns, J., Stent, G. S. and Watson, J. D., eds), pp. 33–42. Cold Spring Harbor Laboratory of Quantitative Biology, New York

Index